Moto Guzzi: A Century of Innovation, Passion, and Heritage

Etienne Psaila

Moto Guzzi: A Century of Innovation, Passion, and Heritage

First Edition: **January 2025**

ISBN: 978-1-923393-87-5

Table of Contents

Chapter 1: Introduction

The Allure of Moto Guzzi

In the vast tapestry of motorcycle history, few names resonate with the same enduring charm and unwavering dedication as Moto Guzzi. Founded in the early 20th century, this illustrious Italian brand has not merely manufactured motorcycles; it has crafted legends that embody the spirit of innovation, resilience, and passion. The allure of Moto Guzzi lies not only in its distinctive engineering marvels but also in the rich narrative that intertwines with the very fabric of motorcycling culture.

From the bustling streets of Mandello del Lario to the global arenas of motorsport, Moto Guzzi has consistently pushed the boundaries of what motorcycles can achieve. Each model that rolled off their assembly lines carried with it a legacy of craftsmanship, blending aesthetic elegance with mechanical prowess. The company's signature transverse V-twin engines became a hallmark of their design philosophy, offering a unique balance of power and smoothness that set Moto Guzzi apart from its contemporaries.

Beyond the machines themselves, Moto Guzzi's story is one of perseverance through tumultuous times. Surviving economic downturns, world wars, and shifting market dynamics, the brand has maintained its identity and continued to thrive. This steadfastness has cultivated a loyal community of riders and enthusiasts who cherish the brand's heritage and eagerly anticipate its future innovations.

The allure is also deeply personal. For many, Moto Guzzi represents freedom on the open road, the thrill of exploration, and the timeless connection between rider and machine. The brand's motorcycles have been companions in countless journeys, each ride weaving new chapters into their storied legacy. Whether cruising along scenic highways or tackling challenging terrains, Moto Guzzi bikes offer an experience that transcends mere transportation, embodying a way of life.

Purpose and Scope of the Book

This book endeavors to chronicle the illustrious journey of Moto Guzzi, tracing its evolution from a fledgling enterprise to a global icon in the motorcycle industry. Through meticulous research and engaging storytelling, it seeks to illuminate the milestones, challenges, and triumphs that have shaped the company's path over the past century.

The purpose is twofold: to provide a comprehensive historical account for enthusiasts and scholars alike, and to celebrate the enduring legacy of Moto Guzzi by highlighting the innovations and personalities that have defined its character. By delving into the company's founding principles, pivotal moments, and cultural impact, this narrative aims to offer readers a nuanced understanding of what makes Moto Guzzi a beloved and respected name in motorcycling.

Spanning the company's inception in 1921 through its modern-day endeavors, the book covers a broad spectrum of topics. It explores the early visionaries who laid the groundwork, the technological breakthroughs that set new standards, and the strategic decisions that navigated the brand through changing landscapes. Each chapter is dedicated to different eras, providing a chronological framework that contextualizes Moto Guzzi's growth within broader historical and economic trends.

Additionally, the book examines the brand's involvement in motorsport, its influence on motorcycle design, and its role in fostering a vibrant community of riders and collectors. By showcasing iconic models and their development, it highlights the engineering excellence and aesthetic sensibilities that have become synonymous with Moto Guzzi.

Moreover, the narrative delves into the company's adaptive strategies in the face of globalization, digital transformation, and evolving consumer preferences. It addresses how Moto Guzzi has balanced tradition with innovation, ensuring that its storied past informs its future trajectory.

Ultimately, this book aspires to be a definitive resource on Moto Guzzi, capturing the essence of a brand that has not only survived but thrived through nearly a century of change. It invites readers to embark on a journey through time, experiencing the passion and dedication that continue to drive Moto Guzzi forward, cementing its place in the annals of motorcycle history.

Chapter 2: The Founding Years: 1921–1930

Origins of Moto Guzzi

In the aftermath of World War I, Italy found itself at a crossroads, grappling with economic recovery and a burgeoning desire for technological advancement. Amidst this backdrop of reconstruction and innovation, the seeds of Moto Guzzi were sown in the picturesque town of Mandello del Lario, nestled on the eastern shore of Lake Como. This region, already renowned for its industrial prowess, particularly in textile manufacturing, was poised to embrace a new chapter in mechanical engineering and transportation.

The early 1920s marked a period of significant transformation in the automotive and motorcycle industries. The demand for reliable, efficient, and affordable motorized vehicles was rising, fueled by both civilian needs and the remnants of military technology repurposed for civilian use. It was within this dynamic environment that the idea of establishing a motorcycle manufacturing enterprise took root, driven by a vision to blend precision engineering with Italian craftsmanship.

Key Founders and Early Vision

The inception of Moto Guzzi can be attributed to the collaboration of three visionary individuals: Carlo Guzzi, Giorgio Parodi, and Giuseppe Morri. Carlo Guzzi, an engineer by training, possessed a profound understanding of mechanical systems and a passion for innovation. His expertise was complemented by Giorgio Parodi, whose business acumen and organizational skills were instrumental in laying the foundational structures of the company. Giuseppe Morri, a seasoned mechanic, brought practical experience and hands-on knowledge essential for the nascent enterprise.

Carlo Guzzi's dream was not merely to build motorcycles but to create machines that embodied reliability, performance, and aesthetic elegance. He envisioned a company that would push the boundaries of motorcycle engineering, setting new standards in design and functionality. This vision was underpinned by a commitment to quality and a desire to position Moto Guzzi as a symbol of Italian ingenuity on the global stage.

The trio's collaboration was marked by meticulous planning and a shared dedication to excellence. They recognized the potential of Mandello del Lario not only for its industrial infrastructure but also for its strategic location, which facilitated access to key markets and resources. With their combined expertise, they embarked on a mission to establish a brand that would resonate with riders and enthusiasts alike.

First Models and Market Entry

Moto Guzzi's journey began with the production of its first motorcycle, the Moto Guzzi 125, introduced in 1921. This initial model was a testament to the founders' commitment to innovation and quality. Featuring a robust single-cylinder engine, the Moto Guzzi 125 was designed to offer both reliability and performance, catering to the needs of everyday riders as well as those seeking a more spirited ride.

The launch of the Moto Guzzi 125 was met with enthusiasm in the Italian market. Its combination of affordability, durability, and efficient performance made it an attractive option for a wide range of customers. The motorcycle quickly gained a reputation for its smooth handling and dependable engine, setting the stage for Moto Guzzi's growing presence in the competitive motorcycle industry.

Building on the success of their first model, Moto Guzzi expanded its lineup in the subsequent years. The introduction of the Moto Guzzi 500 in 1923 marked a significant milestone, offering increased power and enhanced features that appealed to a broader audience. This period also saw the company investing in research and development, striving to incorporate the latest technological advancements into their motorcycles.

The early market entry strategy was characterized by a focus on quality over quantity. Moto Guzzi prioritized the meticulous crafting of each motorcycle, ensuring that every unit met the highest standards of performance and durability. This approach not only garnered the trust of consumers but also established Moto Guzzi as a reliable and reputable brand in the eyes of distributors and retailers.

Establishing the Brand Identity

As Moto Guzzi began to carve out its niche in the motorcycle industry, establishing a distinct brand identity became paramount. The founders understood that in a market teeming with competitors, differentiation was key to sustained success. To this end, Moto Guzzi emphasized several core principles that would define its brand and set it apart from others.

Firstly, engineering excellence was at the heart of Moto Guzzi's identity. The company's commitment to innovation was evident in its continuous efforts to refine and enhance its motorcycle designs. This dedication to superior engineering not only improved performance but also fostered a culture of excellence within the organization.

Secondly, Italian craftsmanship and aesthetic appeal played a crucial role in shaping Moto Guzzi's brand. The motorcycles were not only engineered for performance but also designed with an eye for beauty and elegance. This blend of form and function resonated with riders who appreciated both the mechanical prowess and the visual allure of their machines.

Furthermore, Moto Guzzi cultivated a sense of community and loyalty among its customers. By fostering strong relationships with riders and enthusiasts, the company built a dedicated following that would support its growth and advocate for its brand. This community-centric approach was instrumental in creating a lasting legacy and a sense of pride among Moto Guzzi owners.

Marketing and branding efforts during these foundational years focused on highlighting the unique attributes of Moto Guzzi motorcycles. Advertisements emphasized the reliability, performance, and stylish design of their bikes, appealing to a diverse clientele ranging from daily commuters to adventure seekers. The establishment of a recognizable logo and consistent branding materials further reinforced Moto Guzzi's presence in the market.

In addition to product-focused strategies, Moto Guzzi invested in building a robust distribution network. Strategic partnerships with dealers and distributors ensured that their motorcycles were accessible to customers across Italy and beyond. This expansion was carefully managed to maintain the quality of service and support that Moto Guzzi promised to its riders.

By the end of the 1920s, Moto Guzzi had firmly established itself as a prominent player in the motorcycle industry. The company's unwavering commitment to engineering excellence, combined with its dedication to craftsmanship and community building, had laid a strong foundation for future growth. As the decade drew to a close, Moto Guzzi was poised to navigate the challenges of the coming years, driven by a legacy of innovation and a passion for motorcycling.

Chapter 3: Building Through Adversity: 1930–1945

The Impact of the Great Depression

The dawn of the 1930s cast a long shadow over economies worldwide, and Italy was no exception. The Great Depression, a seismic economic downturn that began in 1929, reverberated across continents, altering industries and reshaping the lives of millions. For Moto Guzzi, a company still in its formative years, the depression presented both formidable challenges and unforeseen opportunities.

Mandello del Lario, the heart of Moto Guzzi's operations, was a town accustomed to industrial resilience. Yet, the pervasive uncertainty of the global market tested the mettle of the fledgling motorcycle manufacturer. Consumer spending plummeted as people grappled with unemployment and financial instability, leading to a significant decline in demand for non-essential goods, including motorcycles. The once bustling assembly lines saw a slowdown, and the company faced the daunting task of maintaining production levels amidst dwindling sales.

However, adversity often breeds innovation, and Moto Guzzi was no stranger to this truth. The economic constraints necessitated a reevaluation of strategies, prompting the company to adopt more efficient production methods and explore cost-effective materials without compromising quality. This period of introspection and adaptation laid the groundwork for Moto Guzzi's enduring commitment to excellence and resourcefulness.

Moreover, the Great Depression underscored the importance of diversification. Recognizing the vulnerability of relying solely on motorcycle sales, Moto Guzzi began to explore alternative revenue streams. This strategic pivot not only provided financial stability but also expanded the company's technical expertise, setting the stage for future innovations.

Innovations Amidst Economic Challenges

In the face of economic adversity, Moto Guzzi demonstrated an unwavering dedication to technological advancement. The constraints of the Great Depression did not dampen the founders' vision; instead, they served as a catalyst for ingenuity. Resource scarcity and the need for cost-effective solutions spurred the company to refine its engineering processes and explore novel designs.

One of the most notable innovations of this era was the enhancement of the Moto Guzzi 500 model. Engineers worked tirelessly to improve engine efficiency, focusing on optimizing fuel consumption without sacrificing performance. This commitment to efficiency was not only a response to economic pressures but also a testament to Moto Guzzi's forward-thinking approach to engineering.

Additionally, the company invested in research and development, fostering a culture of continuous improvement. This period saw the introduction of more streamlined manufacturing techniques, which reduced production costs and minimized waste. These innovations not only bolstered Moto Guzzi's financial standing but also reinforced its reputation for quality and reliability.

The 1930s also witnessed Moto Guzzi's foray into the burgeoning field of electrical engineering. The integration of electrical components into motorcycles was a groundbreaking step, enhancing the functionality and appeal of their models. This pioneering spirit positioned Moto Guzzi at the forefront of technological advancement, distinguishing it from competitors who were slower to embrace change.

Furthermore, the company's commitment to innovation extended beyond engineering. Moto Guzzi began to place greater emphasis on design aesthetics, recognizing that the visual appeal of their motorcycles was as important as their mechanical prowess. This holistic approach to motorcycle design ensured that Moto Guzzi models were not only technically superior but also visually captivating, appealing to a broader spectrum of riders.

Moto Guzzi During World War II

As the world plunged into the turmoil of World War II, Moto Guzzi found itself navigating uncharted waters. The conflict brought about significant disruptions to global supply chains, labor forces, and market dynamics. For Moto Guzzi, the war years were a period of profound transformation, as the company adapted to the demands of a nation at war.

Italy's entry into the war effort necessitated a shift in Moto Guzzi's production priorities. The government's call for industrial support meant that many civilian manufacturing endeavors were repurposed to meet military needs. Moto Guzzi, with its expertise in engineering and manufacturing, was enlisted to contribute to the war effort by producing essential components and machinery for the military.

This transition was not without its challenges. The company had to swiftly reconfigure its production lines, repurposing machinery and retraining workers to meet the new demands. The scarcity of materials further complicated matters, as rationing and resource allocation became central concerns. Despite these hurdles, Moto Guzzi's leadership demonstrated remarkable adaptability, ensuring that production targets were met without compromising the company's standards.

The war years also saw Moto Guzzi's technical capabilities put to the test. The need for robust and reliable machinery led to advancements in engine design and manufacturing processes. These innovations, born out of necessity, would later influence the development of civilian motorcycles, reinforcing Moto Guzzi's reputation for engineering excellence.

Moreover, the war fostered a sense of camaraderie and purpose among Moto Guzzi's workforce. The collective effort to support the nation's war endeavors instilled a deep sense of pride and resilience, traits that would continue to define the company in the post-war era. This period of shared sacrifice and determination solidified Moto Guzzi's identity as a steadfast and dependable manufacturer, capable of overcoming even the most daunting challenges.

Contributions to the War Effort

Moto Guzzi's contributions to the war effort extended beyond manufacturing essential components. The company played a pivotal role in supporting military operations through the provision of motorcycles and specialized vehicles designed for wartime conditions. These motorcycles were integral to various military functions, including reconnaissance, communication, and logistical support.

The design and production of military motorcycles required a balance of durability, reliability, and performance. Moto Guzzi engineers worked tirelessly to develop models that could withstand the harsh conditions of wartime environments. The motorcycles had to be versatile, capable of navigating rugged terrains, and reliable enough to perform consistently under extreme stress. This focus on robustness and adaptability was a testament to Moto Guzzi's engineering prowess and commitment to excellence.

Furthermore, Moto Guzzi collaborated closely with military strategists and engineers to tailor their motorcycles to specific operational needs. This collaboration resulted in the development of specialized features, such as enhanced suspension systems, improved fuel efficiency, and modifications for carrying military equipment. These adaptations not only met the immediate needs of the military but also showcased Moto Guzzi's ability to innovate under pressure.

The company's involvement in the war effort also had significant social implications. Moto Guzzi employed a large workforce, including many women who took on roles traditionally held by men who were serving in the military. This shift not only ensured the continuity of production but also contributed to social changes within the workforce, promoting greater gender inclusivity and diversity.

In addition to manufacturing and design, Moto Guzzi played a role in technological advancements that had lasting impacts beyond the war. The exigencies of wartime production spurred innovations in materials science, manufacturing techniques, and engine technology. These advancements not only enhanced the performance and reliability of military motorcycles but also laid the foundation for post-war civilian models, further cementing Moto Guzzi's legacy as a leader in motorcycle engineering.

The end of World War II brought about a period of reconstruction and reflection for Moto Guzzi. The company emerged from the war with enhanced technical capabilities, a reinforced workforce, and a deeper appreciation for the importance of adaptability and resilience. The experiences and innovations of the war years would shape Moto Guzzi's trajectory in the decades to follow, enabling it to navigate the complexities of the post-war era with confidence and ingenuity.

Conclusion of Chapter 3

The years from 1930 to 1945 were a crucible in which Moto Guzzi's character was forged. The Great Depression tested the company's resilience, compelling it to innovate and adapt in the face of economic hardship. The advent of World War II further challenged Moto Guzzi, necessitating a rapid shift in production priorities and driving technological advancements born out of necessity. Through it all, Moto Guzzi not only survived but emerged stronger, with a reinforced commitment to engineering excellence and a deeper sense of community and purpose.

These formative years underscored the company's ability to navigate adversity with grace and ingenuity, laying a robust foundation for the post-war revival and expansion that would follow. Moto Guzzi's journey through the tumultuous decades of the 1930s and 1940s is a testament to its enduring spirit and unwavering dedication to innovation, qualities that continue to define the brand to this day.

Chapter 4: Post-War Revival and Expansion: 1945–1960

Rebuilding After the War

The conclusion of World War II marked a pivotal turning point for Moto Guzzi, transitioning the company from a wartime manufacturer back to its peacetime roots. Emerging from the shadows of conflict, Moto Guzzi faced the formidable task of rebuilding its operations, restoring its workforce, and reclaiming its position in the civilian motorcycle market. The post-war landscape in Italy was one of both devastation and opportunity, characterized by widespread reconstruction efforts and a burgeoning desire for mobility and freedom among the populace.

Mandello del Lario, the heart of Moto Guzzi's operations, had endured significant strains during the war years. The company's facilities, though adapted for military production, required extensive refurbishment to resume civilian manufacturing. The leadership recognized that restoring production capabilities swiftly was essential not only for economic recovery but also for rekindling the brand's legacy and morale among its employees.

One of the first steps in the rebuilding process was the re-establishment of production lines tailored to consumer motorcycles. This transition involved retooling factories, sourcing civilian-grade materials, and reassembling a workforce that had been stretched thin by wartime demands. The company's commitment to quality remained unwavering, ensuring that the motorcycles produced post-war met the high standards that Moto Guzzi was renowned for.

Economic challenges persisted in the immediate aftermath of the war. Italy was grappling with inflation, shortages of raw materials, and disrupted supply chains. Despite these hurdles, Moto Guzzi leveraged its engineering expertise and strategic planning to navigate the turbulent economic waters. The company focused on optimizing production efficiency, reducing costs without compromising on the quality and performance that defined its motorcycles.

Moreover, the post-war period was a time of innovation and renewal for Moto Guzzi. The experiences and technological advancements gleaned during the war years provided a foundation for future developments. The company harnessed this newfound knowledge to enhance its motorcycle designs, integrating lessons learned from military manufacturing into civilian models. This synergy between past challenges and future aspirations set the stage for Moto Guzzi's resurgence and expansion in the decades to come.

Introduction of Iconic Models

The years following the war were marked by the introduction of several iconic Moto Guzzi models that would cement the brand's reputation for excellence and innovation. These motorcycles not only reflected the company's technical prowess but also embodied the spirit of a society eager to embrace newfound freedoms and mobility.

One of the standout models of this era was the Moto Guzzi Galletto, launched in 1947. This lightweight motorcycle was designed to cater to a wide range of riders, offering a balance of performance, reliability, and affordability. The Galletto quickly became a favorite among both commuters and enthusiasts, thanks to its nimble handling and robust engine. Its success was instrumental in re-establishing Moto Guzzi's presence in the civilian market, signaling the company's return to its core mission of delivering exceptional motorcycles.

Building on the success of the Galletto, Moto Guzzi introduced the Falcone in 1950. This model was a departure from the compact design of the Galletto, featuring a more powerful engine and enhanced suspension systems. The Falcone was engineered for performance, appealing to riders who sought both speed and comfort. Its sophisticated engineering and sleek design made it a symbol of Moto Guzzi's commitment to pushing the boundaries of motorcycle technology.

Another landmark model was the Moto Guzzi Scarab, unveiled in 1954. The Scarab was a trailblazer in several respects, incorporating advanced features such as disc brakes and a fully enclosed chain case. These innovations not only improved the motorcycle's performance and safety but also set new standards for design and functionality in the industry. The Scarab's elegant aesthetics and cutting-edge technology garnered widespread acclaim, further elevating Moto Guzzi's standing among motorcycle manufacturers.

The introduction of these models was complemented by Moto Guzzi's dedication to continuous improvement and customization. The company offered a range of options and accessories, allowing riders to tailor their motorcycles to their individual preferences and needs. This customer-centric approach fostered a loyal following and established Moto Guzzi as a brand that valued both performance and personal expression.

The cumulative impact of these iconic models was profound, laying the groundwork for Moto Guzzi's sustained success and influence in the motorcycle world. Each new release not only showcased the company's technical advancements but also resonated with the aspirations and lifestyles of its riders, reinforcing Moto Guzzi's position as a leader in motorcycle design and engineering.

Expanding into International Markets

With the foundation of robust and innovative motorcycle models, Moto Guzzi set its sights beyond the borders of Italy, embarking on a journey of international expansion that would establish the brand as a global contender in the motorcycle industry. The post-war period was an opportune time for such expansion, as economic recovery efforts and increasing global mobility created a ripe environment for entering new markets.

Moto Guzzi's initial foray into international markets focused on neighboring European countries, leveraging Italy's strong trade relationships and the region's shared cultural affinities. The company established a network of distributors and dealers across Europe, ensuring that its motorcycles were accessible to a diverse clientele. Strategic participation in international motorcycle exhibitions and trade fairs further enhanced Moto Guzzi's visibility and reputation abroad.

The North American market presented a particularly attractive opportunity for Moto Guzzi. The United States, with its vast landscape and car-centric culture, was witnessing a growing interest in motorcycles as symbols of freedom and adventure. Moto Guzzi capitalized on this trend by tailoring its marketing strategies to appeal to American riders. The introduction of models like the Falcone and Scarab in the United States was met with enthusiasm, as these motorcycles embodied the performance and reliability that American consumers sought.

Expanding into the Americas also required Moto Guzzi to navigate logistical challenges, such as establishing reliable distribution channels and providing adequate after-sales support. The company invested in building a robust infrastructure, including service centers and parts distribution networks, to ensure that international customers received the same level of quality and support as their domestic counterparts. This commitment to excellence in international operations fostered trust and loyalty among global riders.

Beyond Europe and North America, Moto Guzzi also explored opportunities in emerging markets across Asia and South America. These regions offered untapped potential, with growing middle classes and increasing demand for motorcycles as affordable and efficient modes of transportation. Moto Guzzi's ability to adapt its models and marketing strategies to cater to the unique preferences and needs of these diverse markets was key to its successful international expansion.

The establishment of Moto Guzzi's presence in international markets not only boosted the company's sales and revenue but also enriched its brand identity. By appealing to a global audience, Moto Guzzi became a symbol of Italian engineering and craftsmanship worldwide, garnering admiration and respect from riders and industry peers alike. This international recognition laid the groundwork for future collaborations, partnerships, and innovations that would further propel Moto Guzzi onto the global stage.

Technological Advancements

The period from 1945 to 1960 was a golden era of technological advancement for Moto Guzzi, characterized by groundbreaking innovations and a relentless pursuit of engineering excellence. These advancements not only enhanced the performance and reliability of Moto Guzzi motorcycles but also set new benchmarks in the industry, reinforcing the brand's reputation as a pioneer in motorcycle technology.

One of the most significant technological milestones of this era was the development of the OHC (Overhead Camshaft) engine. Moto Guzzi engineers, driven by the desire to improve engine performance and efficiency, pioneered the implementation of OHC designs in their motorcycles. This innovation allowed for more precise control of valve timing and improved airflow within the engine, resulting in enhanced power output and smoother operation. The OHC engine became a defining feature of many Moto Guzzi models, distinguishing them from competitors and setting a standard for future engine designs.

In addition to engine advancements, Moto Guzzi made significant strides in suspension and chassis design. The introduction of telescopic front forks and advanced rear suspension systems improved ride comfort and handling, allowing motorcycles to navigate diverse terrains with greater ease and stability. These enhancements not only elevated the riding experience but also expanded the versatility and appeal of Moto Guzzi motorcycles, making them suitable for a wider range of applications and environments.

Electronics and instrumentation also saw notable advancements during this period. Moto Guzzi integrated sophisticated electrical systems into their motorcycles, including improved lighting, ignition systems, and instrumentation panels. These enhancements not only improved safety and functionality but also provided riders with more comprehensive information about their motorcycles' performance and status. The incorporation of advanced electronics underscored Moto Guzzi's commitment to innovation and its ability to anticipate and meet the evolving needs of riders.

Another area of significant technological progress was materials engineering. Moto Guzzi invested in research to develop and utilize lightweight, high-strength materials in their motorcycle construction. The use of advanced alloys and composites reduced the overall weight of the motorcycles while maintaining structural integrity and durability. This focus on materials innovation contributed to improved performance, fuel efficiency, and maneuverability, further enhancing the competitive edge of Moto Guzzi motorcycles.

The company also embraced advancements in manufacturing technology, adopting automated processes and precision engineering techniques to ensure consistent quality and efficiency in production. These technological enhancements streamlined manufacturing operations, reduced production costs, and minimized the margin for error, allowing Moto Guzzi to produce high-quality motorcycles at a scale that met growing demand.

Furthermore, Moto Guzzi's dedication to research and development fostered a culture of continuous improvement and experimentation. Engineers and designers were encouraged to explore new ideas and push the boundaries of what was possible in motorcycle engineering. This culture of innovation led to the creation of experimental models and prototypes that tested the limits of technology and design, many of which influenced the development of future production models.

The cumulative effect of these technological advancements was profound, propelling Moto Guzzi to the forefront of the motorcycle industry. The company's unwavering commitment to innovation not only enhanced the performance and appeal of its motorcycles but also inspired other manufacturers to pursue similar paths of technological excellence. Moto Guzzi's legacy of engineering prowess, established during this transformative period, continues to influence the brand's identity and operations to this day.

Conclusion of Chapter 4

The post-war years from 1945 to 1960 were a time of remarkable transformation and growth for Moto Guzzi. Emerging from the devastation of World War II, the company demonstrated resilience and adaptability, successfully rebuilding its operations and re-establishing its presence in the civilian motorcycle market. The introduction of iconic models like the Galletto, Falcone, and Scarab not only revitalized Moto Guzzi's lineup but also solidified its reputation for quality and innovation.

International expansion played a crucial role in Moto Guzzi's growth, as the company navigated new markets with strategic foresight and a commitment to excellence. By establishing a global presence, Moto Guzzi transcended its Italian roots to become a respected and admired brand worldwide. This international reach not only boosted the company's financial standing but also enriched its cultural and technological exchange, fostering a global community of Moto Guzzi enthusiasts.

Technological advancements during this period were instrumental in shaping the future of Moto Guzzi. The company's pioneering work in engine design, suspension systems, electronics, and materials engineering set new standards in the industry and underscored its dedication to pushing the boundaries of motorcycle technology. These innovations not only enhanced the performance and reliability of Moto Guzzi motorcycles but also inspired a legacy of engineering excellence that continues to define the brand.

As the decade drew to a close, Moto Guzzi stood as a testament to the power of innovation, resilience, and strategic vision. The foundations laid during these formative years positioned the company for sustained success and growth in the ensuing decades. Moto Guzzi's journey through the post-war era exemplifies the enduring spirit of Italian craftsmanship and engineering, qualities that continue to resonate with riders and enthusiasts around the world.

Chapter 5: Golden Era of Innovation: 1960–1980

Pioneering Engineering Feats

The period from 1960 to 1980 stands as a testament to Moto Guzzi's unwavering commitment to engineering excellence and innovation. During these two decades, the company not only solidified its reputation as a leader in motorcycle manufacturing but also pioneered several engineering advancements that would leave an indelible mark on the industry.

One of the most significant achievements of this era was the development and refinement of the Moto Guzzi V8 engine. Introduced in the mid-1960s, the V8 was a marvel of engineering, boasting eight cylinders arranged in a V configuration. This engine was a departure from the traditional designs prevalent at the time, offering unparalleled power, smoothness, and reliability. The V8 engine's design allowed for superior balance and reduced vibrations, enhancing the overall riding experience. Its introduction was not merely a technical triumph but also a bold statement of Moto Guzzi's ambition to push the boundaries of what motorcycles could achieve.

In addition to the V8, Moto Guzzi made significant strides in enhancing engine performance and efficiency across its range. The company invested heavily in research and development, exploring new materials and manufacturing techniques that would improve engine durability and reduce maintenance requirements. Innovations such as improved fuel injection systems and advanced cooling mechanisms were implemented, ensuring that Moto Guzzi motorcycles remained at the forefront of performance and reliability.

Another noteworthy engineering feat was the advancement of suspension systems. Moto Guzzi introduced state-of-the-art telescopic front forks and innovative rear suspension designs that provided superior handling and comfort. These enhancements allowed motorcycles to navigate diverse terrains with ease, catering to both urban commuters and adventure seekers. The focus on suspension technology not only improved ride quality but also contributed to the longevity and resilience of Moto Guzzi motorcycles, making them a preferred choice among riders who valued both performance and durability.

Furthermore, Moto Guzzi embraced the integration of electronic systems into their motorcycles, a relatively novel concept during this period. The introduction of electronic ignition systems and advanced instrumentation panels provided riders with more accurate and reliable information about their motorcycles' performance. These technological integrations enhanced safety and convenience, aligning with Moto Guzzi's ethos of combining functionality with innovation.

The company's dedication to engineering excellence was also reflected in its manufacturing processes. Moto Guzzi adopted precision engineering techniques and automated production methods that ensured consistency and quality across all models. This commitment to manufacturing precision not only elevated the standard of Moto Guzzi motorcycles but also streamlined production, allowing the company to meet growing demand without compromising on quality.

Moto Guzzi in Motorsport

Moto Guzzi's foray into motorsport during the golden era was both a strategic and passionate endeavor, aimed at showcasing the brand's engineering prowess and enhancing its global reputation. The company's participation in various racing events not only provided a platform to test and demonstrate their technological innovations but also fostered a culture of competitiveness and excellence within the organization.

One of the most notable engagements was Moto Guzzi's participation in the prestigious Isle of Man TT races. The challenging course and high-profile nature of the event made it an ideal arena for Moto Guzzi to demonstrate the capabilities of its motorcycles. Riders piloting Moto Guzzi machines consistently delivered impressive performances, earning accolades and cementing the brand's status in the racing community. These successes translated into increased visibility and prestige for Moto Guzzi, attracting both enthusiasts and aspiring engineers to the brand.

In addition to the Isle of Man, Moto Guzzi competed in various

European road races and endurance events. The company's motorcycles were engineered for durability and performance, traits that were rigorously tested in the grueling conditions of endurance racing. Moto Guzzi's commitment to reliability and speed was evident in their race results, with several podium finishes and championship titles that underscored the brand's competitive edge.

Moto Guzzi also embraced the development of specialized racing motorcycles. The introduction of models tailored specifically for racing, such as the Moto Guzzi Le Mans, showcased the company's ability to adapt and innovate. These racing variants featured lightweight frames, enhanced aerodynamics, and high-performance engines, distinguishing them from their civilian counterparts. The success of these specialized models in racing circuits not only demonstrated Moto Guzzi's engineering capabilities but also provided valuable insights that influenced the design and performance of their consumer motorcycles.

The motorsport engagements had a profound impact on Moto Guzzi's brand image and technological advancements. The rigorous demands of racing environments drove the company to continuously refine and enhance their motorcycles, fostering a culture of innovation and excellence. The lessons learned from motorsport endeavors were seamlessly integrated into production models, resulting in motorcycles that benefitted from the same high-performance standards applied in racing.

Moreover, Moto Guzzi's presence in motorsport fostered a loyal community of riders and enthusiasts who admired the brand's dedication to competition and performance. This community not only supported the company's racing efforts but also contributed to the broader Moto Guzzi culture, enhancing the brand's legacy and influence within the motorcycle world.

Signature Designs and Aesthetics

While engineering innovation was paramount, Moto Guzzi also placed a strong emphasis on the aesthetic appeal and design philosophy of its motorcycles. The golden era was marked by a harmonious blend of form and function, resulting in motorcycles that were not only technologically advanced but also visually striking and elegant.

One of the hallmarks of Moto Guzzi's design during this period was the distinctive use of the transverse V-twin engine. This design not only provided mechanical advantages but also contributed to the unique visual identity of Moto Guzzi motorcycles. The exposed engine layout became a signature element, giving the motorcycles a robust and muscular appearance that set them apart from competitors. This design choice emphasized the brand's engineering focus while also enhancing the aesthetic appeal of the motorcycles.

Moto Guzzi also embraced sleek and aerodynamic lines in their motorcycle designs. The company collaborated with renowned designers and industrial artists to create models that were both functional and beautiful. This collaboration resulted in motorcycles that featured streamlined bodies, integrated lighting systems, and ergonomic seating arrangements, providing a comfortable and stylish riding experience. The attention to detail in design extended to the selection of materials and finishes, with Moto Guzzi motorcycles often adorned with chrome accents, polished surfaces, and high-quality upholstery that exuded sophistication and elegance.

Another significant aspect of Moto Guzzi's design philosophy was the emphasis on rider comfort and ergonomics. The company recognized that a motorcycle's design should cater to the needs of the rider, ensuring ease of use and comfort during long journeys. This led to the development of adjustable handlebars, well-padded seats, and intuitive control layouts that enhanced the overall riding experience. By prioritizing rider comfort, Moto Guzzi not only improved the functionality of their motorcycles but also appealed to a broader range of riders, from daily commuters to long-distance enthusiasts.

Moto Guzzi's commitment to aesthetics was also reflected in their approach to color schemes and customization options. The company offered a variety of color choices and customization options, allowing riders to personalize their motorcycles according to their tastes and preferences. This flexibility in design fostered a sense of ownership and pride among riders, further strengthening the bond between the brand and its customers.

The brand's signature designs extended beyond the motorcycles themselves to encompass a holistic brand identity. Moto Guzzi's showrooms, promotional materials, and merchandise all adhered to a consistent design language that emphasized elegance, innovation, and Italian craftsmanship. This cohesive aesthetic presence reinforced Moto Guzzi's image as a premium and desirable brand, attracting discerning riders who valued both performance and style.

Navigating Market Competition

The golden era of Moto Guzzi was not without its challenges, as the company faced increasing competition from both established manufacturers and emerging brands. Navigating this competitive landscape required strategic foresight, adaptability, and a commitment to maintaining the brand's unique identity and values.

During the 1960s and 1970s, the global motorcycle market experienced significant growth, with numerous manufacturers vying for market share. Brands like Harley-Davidson, BMW, and Triumph were formidable competitors, each with their own strengths and loyal customer bases. In this crowded market, Moto Guzzi had to differentiate itself through a combination of engineering innovation, distinctive design, and superior customer service.

One of the key strategies employed by Moto Guzzi to navigate competition was the continuous introduction of innovative models that set new standards in the industry. By consistently pushing the boundaries of engineering and design, Moto Guzzi ensured that its motorcycles remained relevant and desirable in an ever-evolving market. The introduction of advanced features, such as the V8 engine and integrated electronics, provided Moto Guzzi with a competitive edge, attracting riders who sought cutting-edge technology and performance.

Additionally, Moto Guzzi focused on building strong relationships with its dealer network and customers. The company invested in training programs for dealers, ensuring that they were well-equipped to provide exceptional service and support to customers. This emphasis on customer service fostered loyalty and repeat business, as riders felt valued and well-cared-for by the brand. Positive word-of-mouth and a reputation for reliability and quality further bolstered Moto Guzzi's position in the market.

Marketing and branding efforts also played a crucial role in distinguishing Moto Guzzi from its competitors. The company leveraged its rich heritage and association with motorsport to create a compelling brand narrative that resonated with riders. Promotional campaigns highlighted the brand's engineering achievements, racing successes, and commitment to quality, reinforcing Moto Guzzi's image as a premium and aspirational brand.

Moreover, Moto Guzzi adapted to changing consumer preferences and market trends by diversifying its product range. The company introduced a variety of models catering to different segments, from sporty cruisers to versatile touring motorcycles. This diversification allowed Moto Guzzi to appeal to a wider audience, mitigating the risks associated with market fluctuations and changing consumer demands.

The company also embraced globalization, expanding its presence in international markets to reduce reliance on any single region. By establishing a global footprint, Moto Guzzi was able to tap into emerging markets and capitalize on growth opportunities worldwide. This strategic expansion not only increased sales and revenue but also enhanced the brand's global recognition and prestige.

Despite the intense competition, Moto Guzzi's steadfast commitment to its core values of innovation, quality, and craftsmanship ensured that the brand maintained its unique identity and continued to thrive. The ability to balance tradition with modernity, coupled with strategic adaptability, enabled Moto Guzzi to navigate the competitive landscape successfully, securing its place as a revered and respected name in the motorcycle industry.

Conclusion of Chapter 5

The golden era of Moto Guzzi, spanning from 1960 to 1980, was a period marked by remarkable innovation, competitive prowess, and distinctive design excellence. Pioneering engineering feats, exemplified by the introduction of the V8 engine and advancements in suspension and electronics, showcased Moto Guzzi's dedication to pushing the boundaries of motorcycle technology. Their active participation and success in motorsport not only elevated the brand's reputation but also fostered a culture of competitiveness and excellence within the company.

Signature designs and aesthetics further distinguished Moto Guzzi motorcycles, blending mechanical prowess with visual elegance to create machines that were as beautiful as they were powerful. This harmonious integration of form and function resonated with riders, fostering a loyal and passionate community of enthusiasts.

Navigating a fiercely competitive market required strategic foresight and adaptability, qualities that Moto Guzzi demonstrated through continuous innovation, strong customer relationships, and effective marketing strategies. By maintaining its unique identity and embracing diversification and globalization, Moto Guzzi successfully positioned itself as a leader in the global motorcycle industry.

As the chapter concludes, it is evident that the innovations and strategies employed during this golden era laid a robust foundation for Moto Guzzi's sustained success and enduring legacy. The company's ability to blend tradition with innovation, coupled with its unwavering commitment to quality and excellence, ensured that Moto Guzzi remained a beloved and respected name in the hearts of riders around the world. The legacy of these two decades continues to influence Moto Guzzi's trajectory, inspiring future generations to uphold the brand's storied heritage and pioneering spirit.

Chapter 6: Challenges and Transformation: 1980–2000

Economic Pressures and Industry Changes

The dawn of the 1980s ushered in a period of significant economic turbulence and sweeping changes within the global motorcycle industry. For Moto Guzzi, a brand steeped in tradition and recognized for its engineering excellence, these external pressures necessitated a reevaluation of strategies and operations to navigate the evolving landscape.

The global economy in the early 1980s was characterized by instability, marked by recessions, fluctuating exchange rates, and rising inflation. These macroeconomic factors exerted considerable strain on manufacturers worldwide, compelling companies to adopt more efficient production methods and cost-control measures. Moto Guzzi was not immune to these challenges. The company faced increasing production costs, reduced consumer spending, and heightened competition from both established brands and emerging players in the motorcycle market.

Moreover, the motorcycle industry itself was undergoing transformative shifts. The rise of Japanese manufacturers like Honda, Yamaha, and Kawasaki introduced a new era of innovation and affordability, capturing significant market share with their technologically advanced and competitively priced motorcycles. These companies leveraged economies of scale, efficient manufacturing processes, and aggressive marketing strategies to outpace traditional European manufacturers, including Moto Guzzi.

In response to these economic pressures and industry changes, Moto Guzzi embarked on a series of strategic initiatives aimed at enhancing competitiveness and ensuring sustainability. The company recognized the necessity of modernizing its manufacturing processes to improve efficiency and reduce costs. Investments were made in automation and lean manufacturing techniques, streamlining production lines and minimizing waste. These efforts not only optimized operational efficiency but also allowed Moto Guzzi to better manage resources in the face of economic uncertainty.

Additionally, Moto Guzzi sought to diversify its product offerings to appeal to a broader range of consumers. The introduction of new models that catered to different market segments, from sport bikes to touring motorcycles, was part of a strategic effort to mitigate the risks associated with fluctuating market demands. By expanding its portfolio, Moto Guzzi aimed to capture a wider audience and reduce dependency on any single product line, thereby enhancing financial stability.

The company also intensified its focus on quality and innovation as key differentiators in a crowded and competitive market. Emphasizing superior engineering, reliable performance, and distinctive design, Moto Guzzi sought to maintain its reputation for excellence amidst increasing competition. This commitment to quality served as a cornerstone for building customer loyalty and sustaining the brand's prestigious image.

Despite these proactive measures, the economic landscape of the 1980s posed ongoing challenges. Market fluctuations, changing consumer preferences, and the relentless pace of technological advancements required Moto Guzzi to remain agile and responsive. The company's ability to adapt to these external pressures while preserving its core values and heritage was critical to its survival and continued success during this tumultuous period.

Shifts in Ownership and Management

The 1980s and 1990s were decades of significant transformation for Moto Guzzi, not least due to shifts in ownership and management that reshaped the company's strategic direction and operational dynamics. These changes were both a response to external economic pressures and a reflection of internal aspirations for growth and modernization.

In the early 1980s, Moto Guzzi faced financial difficulties exacerbated by declining sales and mounting debts. The company's traditional business model, which had thrived on craftsmanship and engineering prowess, struggled to compete with the mass-produced motorcycles of Japanese manufacturers. Recognizing the need for substantial financial restructuring, Moto Guzzi sought new investment and leadership to steer the company through its challenges.

In 1981, the financial strain culminated in a significant ownership change. The company was acquired by De Tomaso, an Italian automotive firm known for its sports cars and racing heritage. This acquisition marked the beginning of a new era for Moto Guzzi, as De Tomaso sought to leverage its expertise and resources to revitalize the struggling motorcycle manufacturer. The infusion of capital and the strategic guidance from De Tomaso provided Moto Guzzi with the necessary support to stabilize its operations and embark on a path of recovery.

Under De Tomaso's ownership, Moto Guzzi underwent substantial restructuring. New management practices were implemented, focusing on operational efficiency, cost reduction, and product innovation. The leadership team prioritized the development of new models that aligned with contemporary market trends while retaining the brand's distinctive character. This period saw the introduction of motorcycles that balanced performance, style, and affordability, aimed at regaining market share and attracting a new generation of riders.

In 1983, De Tomaso sold a significant stake in Moto Guzzi to the Italian holding company CIR (Compagnia Industriali Riunite). This shift in ownership brought about further changes in management and strategic focus. CIR's investment enabled Moto Guzzi to expand its research and development efforts, fostering innovation and technological advancement. The collaboration between Moto Guzzi and CIR facilitated the introduction of advanced engineering solutions and modern manufacturing techniques, enhancing the company's competitive edge.

The mid-1980s also witnessed a leadership transition as key executives and engineers departed, seeking opportunities elsewhere or retiring after decades of service. This exodus of experienced personnel posed both challenges and opportunities for Moto Guzzi. On one hand, the loss of seasoned professionals created a gap in expertise and institutional knowledge. On the other hand, it provided an opportunity to bring in fresh talent and new perspectives, injecting vitality and innovation into the company's operations.

In response to these changes, Moto Guzzi focused on talent acquisition and development, aiming to build a dynamic workforce equipped to drive the company forward. Emphasis was placed on fostering a culture of continuous improvement, encouraging creativity, and embracing new ideas. This strategic focus on human capital was instrumental in revitalizing Moto Guzzi's organizational structure and enhancing its capacity for innovation.

The late 1990s marked another significant shift in ownership when Moto Guzzi was acquired by Aprilia, an Italian motorcycle manufacturer known for its sport bikes and racing pedigree. This acquisition created a powerful alliance within the Italian motorcycle industry, combining Aprilia's expertise in high-performance motorcycles with Moto Guzzi's legacy of engineering excellence and distinctive design. The partnership aimed to leverage synergies between the two brands, fostering collaboration and mutual growth.

Under Aprilia's stewardship, Moto Guzzi experienced renewed growth and expansion. The integration of resources and shared technological advancements between the two companies led to the development of new models that combined the best of both brands. This period of collaboration not only enhanced Moto Guzzi's product offerings but also solidified its position in the global motorcycle market.

Throughout these decades of ownership and management shifts, Moto Guzzi demonstrated resilience and adaptability. The ability to navigate changes in ownership while maintaining the brand's core values and heritage was crucial to the company's enduring success. These transformations laid the groundwork for Moto Guzzi's continued evolution, positioning it to meet the challenges of the new millennium with renewed vigor and strategic clarity.

Adapting to Technological Shifts

The latter part of the 20th century was marked by rapid technological advancements that revolutionized the motorcycle industry. For Moto Guzzi, adapting to these technological shifts was imperative to remain competitive and meet the evolving demands of riders. The company's approach to technological adaptation was multifaceted, encompassing engine development, electronics integration, manufacturing processes, and environmental sustainability.

One of the most significant technological shifts during this period was the advent of electronic fuel injection (EFI) systems. Traditional carburetors, once the standard for motorcycle engines, were increasingly being replaced by EFI systems that offered improved fuel efficiency, better throttle response, and reduced emissions. Moto Guzzi recognized the importance of this transition and invested in the development and integration of EFI technology into its motorcycles. The adoption of EFI not only enhanced engine performance but also aligned Moto Guzzi with emerging environmental regulations and consumer preferences for more efficient and cleaner-running engines.

In parallel, the integration of advanced electronics into motorcycles became a key area of focus. Moto Guzzi introduced sophisticated electronic systems, including digital instrumentation, anti-lock braking systems (ABS), and electronic suspension controls. These innovations enhanced the safety, performance, and user experience of Moto Guzzi motorcycles, making them more appealing to a modern audience. The incorporation of electronics also facilitated the development of features such as ride-by-wire throttle systems and advanced diagnostic tools, further elevating the brand's technological profile.

Engine development continued to be a cornerstone of Moto Guzzi's technological strategy. The company invested in research and development to refine its signature transverse V-twin engines, enhancing their performance, reliability, and efficiency. Innovations such as variable valve timing, lightweight materials, and improved cooling systems contributed to the evolution of Moto Guzzi's engine technology. These advancements not only improved the riding experience but also reinforced the brand's reputation for engineering excellence.

Manufacturing processes underwent significant modernization as Moto Guzzi sought to enhance production efficiency and quality. The company adopted computer-aided design (CAD) and computer-aided manufacturing (CAM) technologies, enabling more precise and efficient production methods. Automation and robotics were introduced in key areas of the manufacturing process, reducing human error and increasing consistency across production batches. These technological enhancements streamlined operations, reduced production costs, and allowed Moto Guzzi to respond more swiftly to market demands.

Environmental sustainability emerged as a critical consideration in motorcycle manufacturing during this period. Moto Guzzi committed to reducing its environmental footprint by implementing eco-friendly manufacturing practices and developing motorcycles with lower emissions.

The company explored the use of recyclable materials, energy-efficient production methods, and waste reduction strategies. These efforts not only complied with increasingly stringent environmental regulations but also appealed to environmentally conscious consumers, enhancing Moto Guzzi's market appeal.

The integration of technology extended beyond the motorcycles themselves to encompass the company's overall operations. Moto Guzzi embraced information technology to improve supply chain management, inventory control, and customer relationship management (CRM). The implementation of enterprise resource planning (ERP) systems facilitated better coordination across departments, enhancing operational efficiency and enabling more informed decision-making. These technological integrations supported Moto Guzzi's strategic objectives of growth, efficiency, and customer satisfaction.

Moreover, Moto Guzzi fostered a culture of innovation by encouraging collaboration between engineers, designers, and other stakeholders. Cross-functional teams were established to explore new technologies and develop innovative solutions that could be integrated into future models. This collaborative approach ensured that technological advancements were seamlessly incorporated into the design and functionality of Moto Guzzi motorcycles, maintaining the brand's competitive edge in the market.

As the millennium approached, Moto Guzzi's ability to adapt to technological shifts had positioned the company as a forward-thinking and innovative manufacturer. The company's investments in engine technology, electronics, manufacturing modernization, and environmental sustainability not only enhanced the performance and appeal of its motorcycles but also ensured long-term viability in an increasingly competitive and technologically driven industry. These adaptations laid the foundation for Moto Guzzi's continued evolution and success in the 21st century.

Preserving Brand Legacy

Amidst the challenges and transformations of the 1980s and 1990s, preserving Moto Guzzi's rich heritage and brand legacy remained a paramount concern. The company recognized that its long-standing reputation for quality, engineering excellence, and distinctive design was integral to its identity and continued success. Balancing modernization with tradition was a delicate task, requiring strategic initiatives to honor the brand's history while embracing the future.

Moto Guzzi undertook several initiatives to preserve its legacy, starting with the preservation and celebration of its historical models. The company launched limited edition replicas of iconic motorcycles from its past, allowing enthusiasts to experience the craftsmanship and design that defined Moto Guzzi's early years. These replicas not only served as tributes to the brand's heritage but also appealed to collectors and aficionados who valued the legacy of classic motorcycles.

In addition to preserving historical models, Moto Guzzi invested in documenting and archiving its history. The creation of comprehensive records, including design blueprints, engineering documents, and archival photographs, ensured that the company's rich heritage was meticulously preserved. These archives became valuable resources for future generations, providing insights into the company's evolution and the innovations that shaped its journey.

The company also emphasized the importance of storytelling in preserving its legacy. Through marketing campaigns, publications, and events, Moto Guzzi shared the stories of its founders, engineers, and riders who had contributed to the brand's illustrious history. Highlighting these narratives fostered a deeper connection between the brand and its customers, instilling a sense of pride and loyalty among riders who identified with Moto Guzzi's storied past.

Moto Guzzi further reinforced its legacy by maintaining a consistent design language that reflected its heritage. While embracing modern design trends, the company ensured that its motorcycles retained elements that were synonymous with the brand's identity, such as the transverse V-twin engine and the distinctive styling cues that had become hallmarks of Moto Guzzi's aesthetic. This approach allowed the brand to evolve while staying true to its core principles, appealing to both long-time enthusiasts and new riders.

Engagement with the global Moto Guzzi community was another key aspect of preserving the brand's legacy. The company actively supported rider clubs, organized heritage events, and sponsored motorcycle shows that celebrated Moto Guzzi's history and achievements. These initiatives fostered a sense of community and belonging among Moto Guzzi owners and fans, creating a loyal and passionate network that upheld the brand's legacy.

Furthermore, Moto Guzzi prioritized the training and development of its workforce, ensuring that new generations of engineers, designers, and craftsmen were well-versed in the company's traditions and values. Apprenticeship programs and heritage workshops were established to transmit the skills and knowledge that had been honed over decades of motorcycle manufacturing. By investing in human capital, Moto Guzzi ensured that its legacy of craftsmanship and engineering excellence would endure.

The preservation of Moto Guzzi's legacy was also reflected in the company's commitment to quality and reliability. By maintaining high standards in manufacturing and continuously improving product quality, Moto Guzzi reinforced the trust and loyalty of its customers. This unwavering dedication to excellence was a living embodiment of the brand's heritage, ensuring that each motorcycle produced was a testament to Moto Guzzi's storied past and its vision for the future.

As the year 2000 approached, Moto Guzzi stood at the crossroads of tradition and innovation. The company had successfully navigated economic pressures, ownership changes, and technological shifts while steadfastly preserving its legacy. This delicate balance positioned Moto Guzzi to enter the new millennium with a strong sense of identity, honoring its rich history while embracing the opportunities and challenges of the future.

Conclusion of Chapter 6

The period from 1980 to 2000 was a decade of profound challenges and transformative change for Moto Guzzi. Economic pressures and industry shifts tested the company's resilience, compelling it to adopt strategic initiatives to enhance competitiveness and ensure sustainability. Shifts in ownership and management brought new perspectives and resources, facilitating modernization and innovation while navigating the complexities of a rapidly evolving market.

Adapting to technological shifts was a cornerstone of Moto Guzzi's strategy during these decades. The company embraced advancements in engine technology, electronics, and manufacturing processes, positioning itself as a forward-thinking and innovative manufacturer. These technological adaptations not only improved the performance and appeal of Moto Guzzi motorcycles but also ensured long-term viability in an increasingly competitive and technologically driven industry.

Preserving the brand legacy amidst these changes was equally critical. Moto Guzzi undertook deliberate efforts to honor its rich heritage, maintaining a consistent design language, celebrating historical models, and fostering a loyal community of riders and enthusiasts. This commitment to legacy ensured that Moto Guzzi retained its unique identity and continued to resonate with both long-time fans and new generations of riders.

Throughout these transformative years, Moto Guzzi demonstrated remarkable adaptability and strategic foresight. The company's ability to balance tradition with innovation, coupled with its unwavering dedication to quality and excellence, enabled it to navigate the challenges of the late 20th century successfully. These efforts laid a robust foundation for Moto Guzzi's continued evolution, ensuring that the brand remained a revered and respected name in the motorcycle industry.

As the chapter concludes, it is evident that the decade of 1980 to 2000 was instrumental in shaping the trajectory of Moto Guzzi. The lessons learned and the transformations undertaken during this period not only reinforced the company's resilience but also propelled it towards future growth and innovation. Moto Guzzi's journey through these challenging years is a testament to its enduring spirit and commitment to excellence, qualities that continue to define the brand and inspire its legacy.

Chapter 7: The Modern Era: 2000–Present

Revitalization Strategies

As the turn of the millennium approached, Moto Guzzi stood at a crossroads, navigating the complexities of a rapidly evolving motorcycle industry. The early 2000s presented both challenges and opportunities, prompting the company to embark on a series of revitalization strategies aimed at rejuvenating its brand, expanding its market presence, and reaffirming its commitment to engineering excellence.

Central to Moto Guzzi's revitalization was a renewed focus on brand identity and heritage. The company recognized the importance of leveraging its rich history while simultaneously embracing contemporary trends. This delicate balance was essential in appealing to both long-time enthusiasts and a new generation of riders seeking modern performance and style. To this end, Moto Guzzi initiated campaigns that celebrated its storied past, highlighting iconic models and the brand's enduring legacy, while also showcasing its latest innovations and forward-looking designs.

Investment in research and development played a pivotal role in the company's revitalization efforts. Moto Guzzi committed significant resources to advancing its engineering capabilities, aiming to enhance the performance, reliability, and aesthetics of its motorcycles. This period saw the establishment of state-of-the-art R&D

facilities, where engineers and designers collaborated to push the boundaries of motorcycle technology. The emphasis on innovation was not limited to the mechanical aspects of the motorcycles but extended to design, user experience, and manufacturing processes, ensuring that Moto Guzzi remained at the forefront of the industry.

Strategic partnerships and collaborations were another cornerstone of Moto Guzzi's revitalization strategy. By forging alliances with technology firms, design studios, and global distributors, the company expanded its expertise and market reach. These partnerships facilitated the integration of cutting-edge technologies, such as advanced electronics and digital interfaces, into Moto Guzzi's motorcycles, enhancing their appeal and functionality. Additionally, collaborations with renowned designers and artists infused fresh perspectives into the brand's aesthetic, resulting in motorcycles that were both technically superior and visually captivating.

Moto Guzzi also recognized the importance of expanding its global footprint to sustain growth and competitiveness. The company pursued aggressive international expansion, targeting emerging markets in Asia, South America, and the Middle East. Establishing new dealerships, service centers, and marketing hubs in these regions enabled Moto Guzzi to tap into burgeoning demand and cater to diverse consumer preferences. This global expansion was complemented by localized marketing strategies that resonated with regional audiences, ensuring that Moto Guzzi's brand message was both universal and culturally relevant.

Introduction of Modern Models

The early 21st century witnessed the introduction of several modern models that exemplified Moto Guzzi's commitment to innovation and excellence. These motorcycles were designed to meet the evolving needs of riders, offering a blend of performance, comfort, and style that appealed to a broad spectrum of enthusiasts.

One of the flagship models introduced during this period was the Moto Guzzi V85 TT, launched in 2007. The V85 TT was conceived as an adventure-touring motorcycle, combining the rugged versatility required for off-road excursions with the comfort and reliability needed for long-distance travel. Featuring a robust V-twin engine, advanced suspension systems, and enhanced ergonomics, the V85 TT quickly garnered acclaim for its balanced performance and adaptability. Its success marked Moto Guzzi's foray into the adventure-touring segment, a strategic move that expanded the brand's market appeal and showcased its engineering prowess.

Building on the success of the V85 TT, Moto Guzzi introduced the California line, a series of cruiser motorcycles that emphasized style, comfort, and ease of use. The California models were designed with a focus on smooth handling, relaxed ergonomics, and elegant aesthetics, making them ideal for both urban commuting and leisurely rides along scenic routes. The introduction of the California line reinforced Moto Guzzi's reputation for creating motorcycles that seamlessly blend performance with aesthetic appeal, catering to riders who valued both functionality and style.

In 2012, Moto Guzzi unveiled the Guzzi Urban, a modern interpretation of the classic motorcycle designed specifically for urban environments. The Urban featured a compact design, lightweight frame, and efficient engine, making it well-suited for navigating congested city streets. This model reflected Moto Guzzi's adaptability and responsiveness to changing consumer demands, as urban mobility became increasingly important in the context of global urbanization and environmental sustainability.

The introduction of the Moto Guzzi Norge in 2014 marked another significant milestone in the company's modern lineup. The Norge was developed in collaboration with BMW Motorrad, blending Moto Guzzi's Italian design sensibilities with BMW's technological expertise. This partnership resulted in a motorcycle that featured advanced electronics, such as ABS and traction control, alongside Moto Guzzi's signature transverse V-twin engine. The Norge exemplified Moto Guzzi's ability to integrate cutting-edge technology with traditional craftsmanship, offering riders a sophisticated and reliable machine suited for diverse riding conditions.

In recent years, Moto Guzzi has continued to innovate with models like the Stelvio 1200, an evolution of the V85 TT that incorporates more powerful engines, enhanced electronics, and modern design elements. The Stelvio 1200 caters to riders seeking high performance and advanced features, reflecting Moto Guzzi's ongoing commitment to meeting the dynamic demands of the motorcycle market.

Embracing Sustainability and Innovation

As global awareness of environmental issues and sustainability grew, Moto Guzzi recognized the imperative to integrate eco-friendly practices and sustainable technologies into its operations and product offerings. Embracing sustainability was not only a response to regulatory pressures but also a commitment to responsible manufacturing and innovation.

Moto Guzzi initiated several green initiatives aimed at reducing its environmental footprint. The company invested in energy-efficient manufacturing processes, incorporating renewable energy sources and optimizing resource utilization to minimize waste and emissions. These efforts were complemented by the implementation of stringent quality control measures, ensuring that motorcycles were produced with minimal environmental impact without compromising on performance or reliability.

In the realm of product innovation, Moto Guzzi explored alternative fuel technologies and hybrid systems to enhance the efficiency and sustainability of its motorcycles. Research and development teams worked on integrating electric and hybrid propulsion systems into Moto Guzzi's lineup, experimenting with battery technologies and regenerative braking systems. These initiatives aimed to offer riders more sustainable mobility options while maintaining the performance and character that defined Moto Guzzi's motorcycles.

Moto Guzzi also embraced lightweight materials and advanced manufacturing techniques to improve fuel efficiency and reduce emissions. The use of high-strength alloys and composites in motorcycle construction resulted in lighter, more agile machines that consumed less fuel and emitted fewer pollutants. This focus on materials innovation not only enhanced the environmental sustainability of Moto Guzzi's products but also improved their overall performance and handling.

Beyond manufacturing, Moto Guzzi committed to sustainable business practices throughout its supply chain. The company established partnerships with suppliers who adhered to eco-friendly standards, ensuring that raw materials and components were sourced responsibly. Additionally, Moto Guzzi implemented comprehensive recycling programs and waste management systems, promoting a circular economy approach that prioritized the reuse and recycling of materials.

Incorporating sustainability into its brand ethos, Moto Guzzi launched marketing campaigns that highlighted the company's commitment to environmental responsibility.

These campaigns emphasized the eco-friendly aspects of Moto Guzzi's motorcycles and manufacturing processes, appealing to environmentally conscious consumers and reinforcing the brand's image as a responsible and forward-thinking manufacturer.

Moto Guzzi in the Digital Age

The advent of the digital age brought forth new opportunities and challenges for Moto Guzzi, as technology reshaped the landscape of the motorcycle industry. Embracing digital transformation was crucial for the company to enhance operational efficiency, engage with customers, and innovate in product development.

Moto Guzzi leveraged digital technologies to streamline its manufacturing processes and improve quality control. The integration of Industry 4.0 principles, such as automation, data analytics, and the Internet of Things (IoT), enabled Moto Guzzi to optimize production lines, reduce downtime, and ensure consistent product quality. Smart manufacturing systems allowed for real-time monitoring and data-driven decision-making, enhancing the company's ability to respond swiftly to market demands and operational challenges.

In the realm of product development, Moto Guzzi utilized computer-aided design (CAD) and computer-aided engineering (CAE) tools to accelerate the design and testing phases. These digital tools facilitated the creation of precise and intricate designs, enabling engineers to simulate performance and identify potential issues before physical prototypes were built. The adoption of digital twin technology allowed Moto Guzzi to create virtual replicas of motorcycles, enhancing the accuracy of testing and optimization processes.

Digital innovation also extended to the rider experience, with Moto Guzzi incorporating advanced electronics and connectivity features into its motorcycles. Features such as GPS navigation, Bluetooth connectivity, and smartphone integration provided riders with enhanced functionality and convenience. These digital enhancements not only improved the usability of Moto Guzzi's motorcycles but also aligned the brand with the expectations of modern riders who sought seamless integration of technology into their riding experience.

Moto Guzzi embraced e-commerce and digital marketing strategies to reach a broader audience and engage with customers more effectively. The company developed a robust online presence through its website and social media platforms, offering detailed information about its models, facilitating online sales, and fostering community engagement. Interactive digital content, such as virtual tours, 360-degree motorcycle views, and online configurators, allowed customers to explore and personalize Moto Guzzi motorcycles from the comfort of their homes.

Furthermore, Moto Guzzi invested in digital customer service solutions, including online support portals, live chat services, and virtual assistance tools. These digital customer service enhancements ensured that riders received timely and efficient support, enhancing customer satisfaction and loyalty. The use of data analytics and customer relationship management (CRM) systems enabled Moto Guzzi to gain insights into customer preferences and behaviors, informing product development and marketing strategies.

The company also explored the potential of emerging technologies such as augmented reality (AR) and virtual reality (VR) to enhance the rider experience and product visualization. AR applications allowed riders to access real-time information and navigation overlays, while VR provided immersive experiences for showcasing new models and training programs for technicians. These innovative applications of technology positioned Moto Guzzi as a leader in embracing digital advancements within the motorcycle industry.

In addition to enhancing the customer experience, Moto Guzzi leveraged digital technologies to foster collaboration and innovation within the organization. Digital platforms facilitated communication and knowledge sharing among employees, engineers, and designers, promoting a culture

of continuous improvement and creativity. Virtual collaboration tools and project management software streamlined workflows, enabling teams to work more efficiently and effectively across different departments and geographical locations.

Moto Guzzi's participation in digital communities and online forums also played a significant role in strengthening its brand presence. By engaging with riders and enthusiasts through digital channels, Moto Guzzi fostered a sense of community and belonging, encouraging feedback and collaboration that informed product development and brand strategies. This active digital engagement not only enhanced customer loyalty but also provided valuable insights into market trends and consumer preferences.

As the digital landscape continues to evolve, Moto Guzzi remains committed to embracing new technologies and innovations that enhance its motorcycles and business operations. The company's proactive approach to digital transformation ensures that it remains agile and responsive to the changing demands of the motorcycle industry, positioning Moto Guzzi for sustained success in the digital age.

Conclusion of Chapter 7

The modern era, spanning from 2000 to the present, has been a period of dynamic transformation and revitalization for Moto Guzzi. Faced with the challenges of a competitive global market, evolving consumer preferences, and rapid technological advancements, the company embarked on strategic initiatives that rejuvenated its brand, expanded its market presence, and reinforced its commitment to engineering excellence.

Revitalization strategies centered on blending Moto Guzzi's rich heritage with contemporary innovations, ensuring that the brand remained relevant and appealing to both loyal enthusiasts and new generations of riders. The introduction of modern models like the V85 TT, California line, and Norge showcased Moto Guzzi's ability to innovate while honoring its distinctive design philosophy, balancing performance, comfort, and style to meet diverse riding needs.

Embracing sustainability became a cornerstone of Moto Guzzi's operations, reflecting the company's responsibility towards environmental stewardship and its commitment to developing eco-friendly motorcycles. Investments in alternative fuel technologies, lightweight materials, and sustainable manufacturing practices underscored Moto Guzzi's dedication to reducing its environmental footprint and aligning with global sustainability trends.

The digital age brought forth new opportunities for Moto Guzzi to enhance operational efficiency, engage with customers, and innovate in product development. By integrating advanced digital technologies into manufacturing, design, and the rider experience, Moto Guzzi not only improved the functionality and performance of its motorcycles but also elevated the overall customer experience. Digital transformation initiatives enabled the company to streamline operations, foster collaboration, and build a robust online presence that strengthened its global brand identity.

Throughout these transformative years, Moto Guzzi demonstrated remarkable resilience and adaptability, successfully navigating the complexities of a rapidly changing industry while preserving its core values and legacy. The company's ability to balance tradition with innovation, coupled with its unwavering dedication to quality and excellence, ensured that Moto Guzzi remained a revered and respected name in the motorcycle world.

As Moto Guzzi continues to evolve in the modern era, the foundations laid during these two decades of revitalization, sustainability, and digital innovation position the company for ongoing growth and success. The legacy of Moto Guzzi's modern era is one of forward-thinking strategies, technological prowess, and a steadfast commitment to delivering exceptional motorcycles that embody the spirit of Italian craftsmanship and engineering excellence. This enduring legacy not only honors the company's storied past but also paves the way for a promising future, ensuring that Moto Guzzi remains a beloved and influential brand in the global motorcycle landscape.

Chapter 8: Iconic Models Through the Ages

Moto Guzzi's legacy is intricately woven through the fabric of its most iconic models, each embodying the brand's commitment to innovation, craftsmanship, and distinctive design. These motorcycles not only represent milestones in the company's history but also reflect the evolving tastes and demands of riders across generations. This chapter delves into some of Moto Guzzi's most celebrated models, exploring their development, impact, and enduring appeal.

The V8 Series: Power and Performance

In the annals of motorcycle history, few engines have achieved the legendary status of Moto Guzzi's V8 series. Introduced in the mid-1960s, the V8 engine was a groundbreaking achievement, marking a bold departure from conventional motorcycle engineering. This eight-cylinder powerhouse was meticulously engineered to deliver unparalleled performance, smoothness, and reliability, setting a new benchmark for high-performance motorcycles.

The genesis of the V8 series can be traced back to Moto Guzzi's relentless pursuit of engineering excellence. Carlo Guzzi, one of the company's founding fathers, had always envisioned pushing the boundaries of what motorcycles could achieve. The V8 engine was the embodiment of this vision, combining advanced engineering with Moto Guzzi's signature transverse layout. The engine featured eight cylinders arranged in a V configuration, providing a perfect balance of power and smoothness. This design minimized vibrations, ensuring a refined and comfortable riding experience even at high speeds.

The V8 series quickly garnered attention for its exceptional performance capabilities. The engine's robust power output and responsive acceleration made it a favorite among enthusiasts seeking thrilling rides and impressive speed. Models such as the Moto Guzzi V8 Sport became symbols of power and prestige, attracting a loyal following of riders who appreciated the blend of raw performance and sophisticated engineering.

Beyond performance, the V8 series was also celebrated for its reliability and durability. Moto Guzzi engineers meticulously refined the engine's components, ensuring that each V8 motorcycle could withstand the rigors of both everyday use and demanding racing conditions. This commitment to quality fostered a reputation for dependability, solidifying the V8 series as a cornerstone of Moto Guzzi's product lineup.

The legacy of the V8 series extends beyond its immediate impact on the market. The engineering innovations pioneered in the V8 engine influenced subsequent developments within Moto Guzzi and the broader motorcycle industry. Features such as improved cooling systems, advanced fuel injection, and precision machining techniques became standard practices, enhancing the overall quality and performance of motorcycles across the board.

Even decades after their introduction, the V8 series remains a testament to Moto Guzzi's engineering prowess and dedication to excellence. Vintage V8 models are highly sought after by collectors and enthusiasts, celebrated for their timeless design and enduring performance. The V8 series not only defined an era of power and performance but also cemented Moto Guzzi's position as a leader in motorcycle innovation.

The Breva: Classic Elegance

While Moto Guzzi's V8 series epitomized power and performance, the Breva line showcased the brand's flair for classic elegance and refined design. Introduced in the early 1970s, the Breva models were crafted to appeal to riders who valued style, comfort, and timeless aesthetics without compromising on performance.

The Breva series was born out of Moto Guzzi's desire to create motorcycles that seamlessly blended traditional design elements with modern engineering. Drawing inspiration from classic Italian aesthetics, the Breva models featured sleek lines, polished chrome accents, and ergonomically designed frames that exuded sophistication and grace. The attention to detail in the Breva's design made it a standout in a market increasingly dominated by utilitarian and performance-focused motorcycles.

Under the hood, the Breva series was powered by Moto Guzzi's reliable single and twin-cylinder engines, offering a balanced combination of power and efficiency. These engines were meticulously tuned to deliver smooth acceleration and responsive handling, ensuring that the Breva provided an enjoyable and comfortable ride. The integration of advanced suspension systems and precision-engineered brakes further enhanced the motorcycle's performance, making it well-suited for both urban commuting and leisurely rides on scenic routes.

One of the defining features of the Breva series was its focus on rider comfort and ergonomics. The motorcycles were equipped with well-padded seats, adjustable handlebars, and intuitive control layouts, catering to riders seeking long-distance comfort and ease of use. This emphasis on ergonomics made the Breva an ideal choice for touring enthusiasts who valued both style and substance in their motorcycles.

The Breva's timeless design and exceptional craftsmanship resonated with a broad audience, from seasoned riders to newcomers seeking their first motorcycle. Its elegant appearance and reliable performance made it a symbol of understated luxury, appealing to those who appreciated the finer details and classic aesthetics of Moto Guzzi's craftsmanship.

Over the years, the Breva series evolved to incorporate modern technologies and design enhancements while retaining its classic charm. Moto Guzzi's ability to blend tradition with innovation ensured that the Breva remained relevant and desirable in an ever-changing market. Today, vintage Breva models are cherished by collectors for their enduring beauty and exceptional build quality, while newer iterations continue to attract riders who seek a harmonious balance of elegance and performance.

The California: Touring Excellence

The California line represents Moto Guzzi's commitment to creating motorcycles that excel in touring, offering riders a perfect blend of comfort, reliability, and performance for long-distance journeys. Introduced in the early 2000s, the California models quickly became synonymous with touring excellence, catering to riders who sought the freedom and adventure of the open road without sacrificing comfort or style.

At the heart of the California's appeal is its versatile design, engineered to accommodate both urban commutes and extended road trips. The motorcycle features a spacious, ergonomically designed seat, ample storage options, and a relaxed riding position that reduces fatigue during long rides. These elements collectively ensure that the California provides a comfortable and enjoyable experience, whether navigating through city traffic or cruising along highways.

The California's engine, typically a robust V-twin, delivers a smooth and powerful performance that is ideal for touring.

The engine is engineered for reliability and efficiency, providing ample torque and responsive acceleration to handle varying terrains and riding conditions. Advanced features such as electronic fuel injection, anti-lock braking systems (ABS), and integrated riding modes enhance the motorcycle's performance and safety, making it a dependable companion for long journeys.

One of the standout features of the California line is its comprehensive suite of touring amenities. High-quality windshields, adjustable suspension systems, and advanced instrumentation panels equipped with GPS navigation and connectivity features contribute to a superior riding experience. These amenities not only enhance comfort but also provide riders with the tools and information necessary for safe and efficient travel.

The California's design also emphasizes aesthetic appeal, with sleek lines, premium materials, and stylish accents that reflect Moto Guzzi's dedication to both form and function. The motorcycle's elegant appearance, combined with its practical features, makes it a favorite among riders who value both style and substance in their touring machines.

Over the years, the California line has seen numerous updates and enhancements, incorporating the latest technologies and design innovations to meet the evolving needs of touring enthusiasts. Moto Guzzi's commitment to continuous improvement ensures that the California remains a top choice for riders seeking a reliable and stylish touring motorcycle.

The California's success is a testament to Moto Guzzi's ability to create motorcycles that cater to specific riding styles and preferences. By focusing on the unique demands of touring, Moto Guzzi has carved out a niche for the California line, making it an enduring symbol of excellence in the realm of long-distance riding.

Limited Editions and Special Releases

In addition to its flagship models, Moto Guzzi has a storied tradition of releasing limited edition and special release motorcycles that celebrate the brand's heritage, commemorate significant milestones, and showcase unique design and engineering innovations. These exclusive models are highly anticipated by enthusiasts and collectors alike, offering a unique blend of rarity, craftsmanship, and distinctive features.

Limited editions often draw inspiration from Moto Guzzi's rich history, reimagining classic designs with modern enhancements. For instance, the Moto Guzzi V7 Le Mans homage pays tribute to the brand's racing legacy, incorporating design elements reminiscent of the original racing bikes while integrating contemporary technologies for improved performance and reliability. These homage models allow riders to experience the nostalgia of Moto Guzzi's racing heritage in a modern package, bridging the gap between past and present.

Special releases also serve as platforms for Moto Guzzi to experiment with innovative designs and technologies. These models often feature bespoke elements such as custom paint schemes, unique color combinations, and specialized accessories that set them apart from standard production motorcycles. The attention to detail in these limited editions underscores Moto Guzzi's commitment to craftsmanship and exclusivity, making each special release a coveted addition to any collection.

Collaborations with artists, designers, and other brands have also resulted in unique limited edition models that push the boundaries of motorcycle design. These collaborations infuse fresh perspectives and creative elements into Moto Guzzi's lineup, resulting in motorcycles that are as much works of art as they are engineering marvels. Such partnerships not only enhance the aesthetic appeal of the motorcycles but also broaden Moto Guzzi's appeal to a diverse audience, including art aficionados and design enthusiasts.

One notable example is the Moto Guzzi V9 Volpe, a limited edition model that celebrates the brand's commitment to innovation and distinctive design. Featuring a sleek, fox-inspired silhouette, the V9 Volpe combines aerodynamic efficiency with elegant styling, resulting in a motorcycle that is both visually striking and performance-oriented. This model exemplifies Moto Guzzi's ability to create limited editions that are not only unique and exclusive but also embody the brand's core values of innovation and craftsmanship.

Special releases often come with enhanced performance features and exclusive technologies that are not available in standard models. These enhancements can include upgraded suspension systems, advanced electronic controls, and high-performance engine tuning, offering riders an elevated riding experience. The incorporation of these specialized features further distinguishes limited edition models, making them desirable for riders seeking the pinnacle of performance and exclusivity.

The allure of limited editions and special releases lies in their rarity and the unique stories they tell. Each exclusive model represents a specific moment in Moto Guzzi's history, a celebration of its achievements, or an exploration of new design frontiers. Collectors and enthusiasts cherish these models not only for their technical and aesthetic qualities but also for the rich narratives they embody, preserving Moto Guzzi's legacy through each meticulously crafted motorcycle.

Conclusion of Chapter 8

Moto Guzzi's iconic models are more than just motorcycles; they are embodiments of the brand's enduring commitment to innovation, craftsmanship, and distinctive design. From the powerful and legendary V8 series to the elegant Breva, the touring excellence of the California, and the exclusivity of limited editions and special releases, each model has played a pivotal role in shaping Moto Guzzi's legacy.

These motorcycles reflect the company's ability to adapt to changing market demands while staying true to its core values and heritage. They showcase Moto Guzzi's dedication to engineering excellence, aesthetic sophistication, and rider-centric design, ensuring that each model not only meets but exceeds the expectations of riders and enthusiasts worldwide.

The enduring appeal of these iconic models lies in their ability to capture the essence of Moto Guzzi's spirit—blending tradition with innovation, performance with style, and reliability with exclusivity. As Moto Guzzi continues to evolve in the modern era, its iconic models serve as a testament to the brand's rich history and its unwavering commitment to delivering exceptional motorcycles that inspire passion and loyalty among riders around the globe.

Chapter 9: Moto Guzzi in Motorsport

Motorcycle racing has always been a crucible for innovation, testing the limits of engineering, endurance, and rider skill. For Moto Guzzi, participation in motorsport has been both a proving ground and a platform for showcasing the brand's engineering prowess and competitive spirit. This chapter delves into Moto Guzzi's illustrious journey in the world of motorsport, exploring its early ventures, notable achievements, the profound influence on the brand's image, and the enduring legacy that continues to inspire modern racing endeavors.

Early Racing Ventures

Moto Guzzi's foray into motorsport began almost as soon as the company was founded. In the early years of the 20th century, motorcycle racing was gaining popularity as a way to demonstrate the capabilities of newly emerging motorcycle manufacturers. Recognizing the potential of motorsport to showcase their engineering excellence and enhance brand visibility, Moto Guzzi eagerly embraced racing as a strategic avenue for growth and recognition.

The 1920s and 1930s marked Moto Guzzi's initial steps into the competitive racing scene. The company entered various local and national races, eager to test the limits of their motorcycles and gain valuable insights into performance enhancement. These early racing ventures were characterized by a spirit of experimentation and a relentless pursuit of speed and reliability. Moto Guzzi's engineers meticulously fine-tuned their machines, focusing on engine performance, handling, and durability to withstand the demanding conditions of racing circuits.

One of the pivotal moments in Moto Guzzi's early racing history was participation in the prestigious Mille Miglia, an endurance race that traversed Italy's challenging terrains. The Mille Miglia was not only a test of speed but also of endurance and mechanical resilience. Moto Guzzi's entry into such high-profile events underscored the company's commitment to excellence and provided a platform to demonstrate the robustness and performance of their motorcycles under extreme conditions.

Riders piloting Moto Guzzi motorcycles began to earn accolades for their skill and the machines' reliability. These successes in early races laid the foundation for Moto Guzzi's reputation as a manufacturer capable of producing competitive and dependable racing motorcycles. The experiences and lessons gleaned from these ventures were instrumental in refining Moto Guzzi's engineering practices and enhancing the overall quality of their production models.

Achievements and Championships

Moto Guzzi's dedication to motorsport bore significant fruit in the ensuing decades, with the company achieving numerous victories and championships that cemented its status as a formidable competitor in the racing world. These achievements were not only a testament to the brand's engineering capabilities but also to the skill and bravery of its riders.

One of Moto Guzzi's most notable successes came in the realm of Grand Prix racing. Throughout the 1950s and 1960s, Moto Guzzi dominated various Grand Prix events, particularly in the 500cc class. The company's motorcycles, renowned for their powerful V-twin engines and superior handling, proved to be formidable contenders against the rising dominance of Japanese manufacturers. Moto Guzzi's triumphs in these high-stakes competitions showcased the brand's ability to compete at the highest levels of motorcycle racing.

Endurance racing was another arena where Moto Guzzi left an indelible mark. The company's participation in the 24 Hours of Le Mans and the Bol d'Or races highlighted their expertise in building motorcycles that could sustain performance over extended periods. The Bol d'Or, in particular, became synonymous with Moto Guzzi's endurance capabilities, with the company securing multiple victories that underscored the reliability and durability of their motorcycles.

Moto Guzzi also excelled in production-based racing, where motorcycles that closely mirrored their commercial counterparts competed in various national and international events. These races provided an opportunity to demonstrate the performance of Moto Guzzi's production models, bridging the gap between racing technology and consumer motorcycles. The success of Moto Guzzi in production racing events not only boosted sales but also enhanced the brand's reputation for producing high-performance, reliable motorcycles.

The pinnacle of Moto Guzzi's racing achievements was perhaps their success in the Isle of Man TT races. Known for their grueling courses and demanding conditions, the Isle of Man TT races were a true test of a motorcycle's performance and a rider's skill. Moto Guzzi's victories in these races were celebrated as triumphs of both engineering and human endurance, further solidifying the brand's legacy in the motorsport arena.

Influence on Brand Image

Moto Guzzi's successes in motorsport had a profound impact on the brand's image and market positioning. Racing victories served as powerful endorsements of the company's engineering excellence and the performance capabilities of its motorcycles. These achievements not only garnered respect and admiration within the racing community but also resonated with consumers, enhancing Moto Guzzi's reputation as a manufacturer of high-quality, high-performance motorcycles.

The association with motorsport imbued Moto Guzzi with an aura of prestige and reliability. Riders and enthusiasts began to view Moto Guzzi motorcycles as symbols of speed, endurance, and technological prowess. This perception translated into increased brand loyalty and a stronger emotional connection between the company and its customers. The thrill of racing victories and the stories of daring riders became integral to Moto Guzzi's brand narrative, reinforcing the company's identity as a leader in motorcycle innovation and performance.

Furthermore, Moto Guzzi's involvement in motorsport provided invaluable marketing and promotional opportunities. High-profile racing events attracted media attention and public interest, offering Moto Guzzi a platform to showcase their motorcycles to a wider audience. The visual spectacle of racing, coupled with the drama of competition, created compelling narratives that enhanced the brand's visibility and appeal. Advertising campaigns that highlighted Moto Guzzi's racing successes resonated with consumers, positioning the brand as aspirational and desirable.

The technological advancements driven by motorsport participation also played a crucial role in shaping Moto Guzzi's brand image. Innovations developed for racing were often adapted and integrated into commercial models, demonstrating Moto Guzzi's commitment to continuous improvement and excellence. This synergy between racing technology and consumer motorcycles underscored the brand's dedication to delivering cutting-edge performance and reliability, further enhancing its reputation for engineering excellence.

Moto Guzzi's motorsport legacy also fostered a sense of community and pride among riders and enthusiasts. The shared passion for racing and the celebration of Moto Guzzi's achievements created a loyal and engaged customer base that not only supported the brand's racing endeavors but also advocated for its motorcycles. This community-centric approach strengthened Moto Guzzi's brand image, fostering a deep sense of belonging and pride among its riders.

Legacy in Modern Motorsports

The legacy of Moto Guzzi's involvement in motorsport continues to influence the brand's presence and initiatives in modern racing arenas. While the landscape of motorcycle racing has evolved with advancements in technology and changes in competition dynamics, Moto Guzzi's foundational experiences and achievements remain integral to its identity and strategic direction.

In contemporary motorsport, Moto Guzzi maintains a presence in various racing disciplines, including endurance racing and production-based competitions. The company continues to participate in prestigious events, leveraging its rich heritage to inspire current and future generations of riders and engineers. Modern Moto Guzzi racing teams embody the same spirit of innovation and excellence that characterized the brand's early racing ventures, striving to achieve new milestones and uphold the legacy of past successes.

Moto Guzzi's influence extends beyond the racetrack, impacting motorcycle design and engineering in the broader industry. The company's commitment to performance and reliability, honed through decades of racing, continues to drive technological advancements and inspire innovation within Moto Guzzi's product development teams. Features such as advanced suspension systems, high-performance engines, and integrated electronics, originally developed for racing, are now standard in Moto Guzzi's commercial models, reflecting the enduring impact of motorsport on the brand's engineering philosophy.

Additionally, Moto Guzzi fosters a strong connection between its racing heritage and its consumer base through various initiatives and events. The company organizes and sponsors racing events, rider training programs, and community gatherings that celebrate its motorsport legacy. These initiatives not only honor Moto Guzzi's past achievements but also encourage active participation in modern racing, ensuring that the brand's motorsport spirit remains vibrant and relevant.

Moto Guzzi also collaborates with contemporary racing organizations and technology partners to stay at the forefront of motorsport innovation. These collaborations facilitate the exchange of knowledge and expertise, enabling Moto Guzzi to integrate the latest advancements into its racing and commercial motorcycles. By embracing modern technologies and fostering partnerships within the racing community, Moto Guzzi ensures that its motorsport legacy continues to evolve and inspire future innovations.

The legacy of Moto Guzzi in motorsport is also preserved through archival efforts and historical commemorations. The company maintains detailed records of its racing history, celebrating significant victories and honoring the contributions of legendary riders and engineers. Exhibitions, museums, and digital archives showcase Moto Guzzi's motorsport heritage, providing enthusiasts and the general public with insights into the brand's racing journey and its enduring impact on motorcycle racing.

In essence, Moto Guzzi's legacy in motorsport is a testament to the brand's enduring commitment to excellence, innovation, and competitive spirit. As Moto Guzzi continues to navigate the modern racing landscape, the lessons and achievements of the past guide its path forward, ensuring that the brand remains a revered and influential presence in the world of motorcycle racing.

Conclusion of Chapter 9

Moto Guzzi's journey in motorsport is a narrative of passion, innovation, and relentless pursuit of excellence. From its early racing ventures that laid the groundwork for a storied legacy, to its numerous achievements and championships that solidified its status as a formidable competitor, Moto Guzzi has consistently demonstrated its commitment to engineering prowess and competitive spirit. The influence of motorsport on the brand's image has been profound, elevating Moto Guzzi to a position of prestige and reliability in the eyes of riders and enthusiasts worldwide.

The legacy of Moto Guzzi in modern motorsports continues to inspire and drive the brand's initiatives, ensuring that the spirit of competition and innovation remains at the core of its identity. By maintaining a presence in contemporary racing disciplines, fostering technological advancements, and preserving its rich motorsport heritage, Moto Guzzi honors its past while forging a path towards future achievements.

As Moto Guzzi continues to evolve, its motorsport legacy serves as both a foundation and a beacon, guiding the company towards new horizons of performance, design, and rider satisfaction. The enduring impact of Moto Guzzi's involvement in motorsport not only celebrates the brand's historical achievements but also fuels its ongoing quest for excellence in the dynamic and ever-evolving world of motorcycle racing.

Chapter 10: Technological Innovations and Engineering Marvels

Moto Guzzi has long been celebrated not only for its rich heritage and distinctive designs but also for its unwavering commitment to technological innovation and engineering excellence. Over the decades, the company has pioneered numerous advancements that have not only enhanced the performance and reliability of its motorcycles but have also set new standards within the industry. This chapter explores the key technological innovations and engineering marvels that have defined Moto Guzzi's journey, focusing on engine developments, suspension and frame innovations, electronic enhancements, and sustainable technologies.

Engine Developments

At the heart of every Moto Guzzi motorcycle lies its engine, a testament to the brand's dedication to engineering excellence and performance. Throughout its history, Moto Guzzi has consistently pushed the boundaries of engine design, developing powerplants that deliver both performance and reliability.

One of the most significant milestones in Moto Guzzi's engine development was the introduction of the transverse V-twin engine. This distinctive configuration, with cylinders arranged perpendicular to the frame, not only gave Moto Guzzi motorcycles their unique appearance but also provided mechanical advantages. The transverse layout allowed for better cooling and reduced vibrations, resulting in a smoother and more balanced ride. This engine design became a hallmark of Moto Guzzi, setting the brand apart from its competitors and becoming synonymous with its identity.

In the 1960s, Moto Guzzi took a bold step forward with the development of the V8 engine. This eight-cylinder marvel was a pioneering achievement in motorcycle engineering, offering unparalleled power and smoothness. The V8 engine featured a complex design with eight cylinders arranged in a V configuration, delivering exceptional performance and a distinctive sound that captivated enthusiasts. Despite its complexity, the V8 proved to be remarkably reliable, showcasing Moto Guzzi's ability to innovate without compromising on durability.

Throughout the 1980s and 1990s, Moto Guzzi continued to refine its engine technology, introducing advancements such as electronic fuel injection (EFI) systems. EFI replaced traditional carburetors, offering improved fuel efficiency, better throttle response, and reduced emissions. This transition not only enhanced the performance of Moto Guzzi's engines but also aligned the brand with evolving environmental standards and consumer expectations.

In recent years, Moto Guzzi has embraced modern engineering techniques to further enhance engine performance and efficiency. Innovations such as variable valve timing, lightweight materials, and advanced cooling systems have been integrated into their engines, resulting in powerplants that are both powerful and efficient. These advancements reflect Moto Guzzi's commitment to staying at the forefront of engine technology, ensuring that their motorcycles continue to deliver exceptional performance while meeting contemporary demands for efficiency and sustainability.

Suspension and Frame Innovations

Moto Guzzi's dedication to delivering a superior riding experience extends beyond engine performance to encompass suspension and frame design. The company has consistently sought to enhance the handling, comfort, and stability of its motorcycles through innovative suspension systems and robust frame constructions.

In the early years, Moto Guzzi introduced telescopic front forks, a significant improvement over the traditional rigid front ends used by many manufacturers. This advancement provided better shock absorption and improved handling, allowing riders to navigate uneven terrains with greater ease and comfort. The adoption of telescopic forks was a key factor in enhancing the overall ride quality of Moto Guzzi motorcycles, making them more versatile and enjoyable across a variety of riding conditions.

The development of advanced rear suspension systems marked another milestone in Moto Guzzi's engineering journey. The company experimented with different suspension configurations to achieve the optimal balance between comfort and performance. Innovations such as mono-shock and dual-shock setups were explored, each offering distinct advantages in terms of ride stability and handling dynamics. These advancements allowed Moto Guzzi to tailor its suspension systems to the specific needs of different motorcycle models, ensuring that each bike delivered a riding experience that matched its intended purpose.

Moto Guzzi also invested in the development of lightweight yet strong frame materials, utilizing high-strength alloys and composites to reduce overall motorcycle weight without compromising structural integrity. The use of these materials contributed to improved handling and maneuverability, making Moto Guzzi motorcycles more agile and responsive. Additionally, advanced frame designs, incorporating elements such as trellis structures and perimeter frames, provided enhanced rigidity and stability, further elevating the performance capabilities of Moto Guzzi's motorcycles.

In the pursuit of perfection, Moto Guzzi continuously refined its suspension and frame technologies, integrating feedback from riders and leveraging advancements in materials science and engineering. The result has been a lineup of motorcycles that offer exceptional handling, superior comfort, and robust performance, cementing Moto Guzzi's reputation for engineering excellence and rider-centric design.

Electronic Enhancements

The integration of advanced electronics into motorcycles has revolutionized the riding experience, and Moto Guzzi has been at the forefront of this technological evolution. By embracing electronic enhancements, Moto Guzzi has enhanced the functionality, safety, and overall performance of its motorcycles, aligning with the digital age's demands and rider expectations.

One of the most notable electronic advancements in Moto Guzzi's lineup is the introduction of anti-lock braking systems (ABS). ABS technology prevents wheel lock-up during sudden braking, maintaining traction and stability, and significantly reducing the risk of accidents. The incorporation of ABS into Moto Guzzi's motorcycles has enhanced rider safety, providing greater confidence and control in various riding conditions.

Electronic fuel injection (EFI) systems, mentioned earlier in the context of engine developments, have also played a crucial role in enhancing motorcycle performance and efficiency. EFI offers precise control over fuel delivery, improving combustion efficiency, reducing emissions, and ensuring optimal engine performance. This technology has become a standard feature in Moto Guzzi's modern engines, contributing to their reliability and environmental friendliness.

Moto Guzzi has also embraced advanced instrumentation and rider interfaces, integrating digital displays and connectivity features into their motorcycles. These enhancements provide riders with real-time information about engine performance, fuel levels, navigation, and other critical metrics, enhancing the overall riding experience. Features such as Bluetooth connectivity allow riders to connect their smartphones to the motorcycle's systems, enabling access to music, calls, and navigation directly from the handlebars.

Ride-by-wire throttle systems represent another significant electronic enhancement in Moto Guzzi's motorcycles. This technology replaces traditional mechanical throttle cables with electronic controls, allowing for more precise throttle response and customizable riding modes. Riders can switch between different performance settings, such as eco, sport, and touring modes, tailoring the motorcycle's behavior to their preferences and riding conditions.

Furthermore, Moto Guzzi has integrated advanced suspension controls, such as electronic suspension damping adjustments, which allow riders to fine-tune the suspension settings based on terrain and riding style. These electronic enhancements not only improve ride comfort and handling but also showcase Moto Guzzi's commitment to leveraging technology to enhance the rider's experience.

Incorporating these electronic advancements has positioned Moto Guzzi as a forward-thinking brand, embracing modern technologies to deliver motorcycles that are not only powerful and stylish but also safe, efficient, and connected. These enhancements have broadened the appeal of Moto Guzzi's motorcycles, attracting a diverse range of riders who seek the benefits of advanced technology integrated seamlessly into their motorcycles.

Sustainable Technologies

In an era where environmental sustainability is paramount, Moto Guzzi has taken proactive steps to integrate eco-friendly technologies into its motorcycles and manufacturing processes. The company's commitment to sustainability reflects its responsibility towards the environment and its dedication to innovation that benefits both riders and the planet.

One of the primary areas of focus for Moto Guzzi has been the development of fuel-efficient engines that reduce emissions and improve overall environmental performance. By optimizing engine design and implementing technologies such as electronic fuel injection (EFI) and variable valve timing, Moto Guzzi has significantly enhanced fuel efficiency and minimized harmful emissions. These advancements not only comply with increasingly stringent environmental regulations but also cater to environmentally conscious consumers who prioritize sustainability in their purchasing decisions.

In addition to engine efficiency, Moto Guzzi has explored alternative propulsion technologies, including hybrid and electric powertrains. The company has invested in research and development to create motorcycles that incorporate electric motors alongside traditional combustion engines, offering riders a hybrid option that combines the best of both worlds. These hybrid models provide reduced emissions and improved fuel economy without sacrificing performance, aligning with global trends towards greener transportation solutions.

Moto Guzzi has also embraced the use of sustainable materials in motorcycle construction. The company sources high-strength, lightweight alloys and composites that not only enhance performance and handling but also reduce the environmental impact of manufacturing. Additionally, Moto Guzzi has implemented recycling programs and waste reduction initiatives within its production facilities, ensuring that materials are reused and waste is minimized. These efforts contribute to a more sustainable manufacturing process, reducing the company's carbon footprint and promoting responsible resource management.

Another significant aspect of Moto Guzzi's sustainability initiatives is the adoption of energy-efficient manufacturing practices. The company has invested in renewable energy sources, such as solar and wind power, to supply its production facilities, reducing reliance on fossil fuels and lowering greenhouse gas emissions. Energy-efficient machinery and automation systems have been integrated into the manufacturing process, enhancing productivity while minimizing energy consumption and operational costs.

Moto Guzzi also emphasizes the importance of sustainability in its product lifecycle. From design and manufacturing to maintenance and disposal, the company implements practices that promote environmental responsibility. Motorcycles are designed for longevity and ease of maintenance, reducing the need for frequent replacements and minimizing waste. Additionally, Moto Guzzi offers recycling programs for old motorcycles and components, ensuring that materials are repurposed and reducing the environmental impact of disposal.

The company's commitment to sustainability extends beyond its products and operations to encompass broader community and environmental initiatives. Moto Guzzi participates in environmental conservation projects, supports sustainable mobility programs, and engages in partnerships with organizations dedicated to preserving natural resources and promoting eco-friendly transportation solutions. These initiatives reflect Moto Guzzi's holistic approach to sustainability, integrating environmental responsibility into every aspect of its business.

Conclusion of Chapter 10

Moto Guzzi's journey through technological innovation and engineering marvels underscores the brand's enduring commitment to excellence, performance, and sustainability. From pioneering engine developments and advanced suspension systems to integrating cutting-edge electronics and embracing sustainable technologies, Moto Guzzi has consistently pushed the boundaries of what is possible in motorcycle engineering.

The company's ability to blend tradition with innovation has allowed it to create motorcycles that not only honor its rich heritage but also meet the evolving demands of modern riders. Engine advancements have provided unparalleled performance and reliability, while suspension and frame innovations have enhanced handling and comfort. Electronic enhancements have elevated the rider's experience, offering safety, connectivity, and customization options that align with contemporary technological trends. Meanwhile, sustainable technologies reflect Moto Guzzi's responsibility towards the environment, ensuring that the brand's legacy is one of both innovation and stewardship.

As Moto Guzzi continues to evolve, its focus on technological advancements and engineering excellence remains steadfast. The company's dedication to integrating the latest technologies while preserving its distinctive design philosophy ensures that Moto Guzzi motorcycles remain at the forefront of the industry, admired for their performance, reliability, and timeless elegance.

Chapter 11: Cultural Impact and Community

Moto Guzzi's influence extends far beyond the realm of motorcycle manufacturing; it has woven itself into the cultural and social fabric of communities around the world. The brand's enduring legacy is reflected not only in its engineering achievements and iconic models but also in its profound impact on popular culture, the vibrant communities of riders and enthusiasts it has fostered, the festivals and events that celebrate its heritage, and the dedicated efforts to preserve its rich history through museums and archival endeavors. This chapter explores the multifaceted cultural impact and the strong sense of community that Moto Guzzi has cultivated over the decades.

Moto Guzzi in Popular Culture

Moto Guzzi motorcycles have long been symbols of freedom, adventure, and Italian craftsmanship, making significant appearances in various facets of popular culture. From films and literature to music and art, Moto Guzzi has served as both inspiration and icon, embodying the spirit of the open road and the allure of mechanical beauty.

In cinema, Moto Guzzi motorcycles have featured prominently in numerous films, often representing the epitome of style and performance. Classic Italian films of the mid-20th century showcased Moto Guzzi bikes as the preferred choice for protagonists seeking both reliability and sophistication. These cinematic portrayals not only highlighted the motorcycles' aesthetic appeal but also reinforced their status as aspirational machines. In more contemporary films, Moto Guzzi continues to make appearances, often symbolizing a blend of heritage and modernity that resonates with diverse audiences.

Literature and art have also embraced Moto Guzzi, with numerous authors and artists drawing inspiration from the brand's rich history and distinctive designs. Novels and autobiographies of famous riders and motorcycle enthusiasts frequently feature Moto Guzzi motorcycles, using them as symbols of personal freedom and the pursuit of adventure. In the visual arts, Moto Guzzi's elegant lines and iconic silhouettes have been the subjects of paintings, sculptures, and photographic exhibitions, celebrating the brand's contribution to automotive aesthetics.

Music has not been left untouched by Moto Guzzi's influence. Several musicians and bands have referenced Moto Guzzi in their lyrics and album art, using the motorcycles as metaphors for freedom, rebellion, and the open road. The rhythmic roar of Moto Guzzi engines has inspired countless songs that capture the essence of the riding experience, further embedding the brand into the cultural zeitgeist.

Moreover, Moto Guzzi's distinctive design and engineering have made it a muse for designers and engineers beyond the motorcycle industry. The brand's ability to blend form and function seamlessly has influenced automotive design trends and inspired innovations in other fields of engineering and craftsmanship. This cross-pollination of ideas underscores Moto Guzzi's role as a cultural and technological influencer, shaping not only motorcycles but also broader design philosophies.

Rider Communities and Clubs

Central to Moto Guzzi's cultural impact is the vibrant and dedicated community of riders and enthusiasts that the brand has cultivated over the years. These communities and clubs are more than just groups of motorcycle owners; they are tight-knit networks that foster camaraderie, shared passion, and mutual support among members.

Moto Guzzi clubs can be found in virtually every corner of the globe, each with its own unique identity and activities. These clubs organize regular rides, social gatherings, and events that bring together riders to celebrate their shared love for Moto Guzzi motorcycles. Whether cruising through scenic landscapes, participating in charity rides, or engaging in spirited discussions about the latest models and modifications, these communities provide a sense of belonging and purpose for their members.

The camaraderie within Moto Guzzi communities extends beyond the riders themselves. Families of riders often participate in club activities, creating a multigenerational bond centered around the brand. This inclusive approach ensures that Moto Guzzi's legacy is passed down through generations, fostering a deep-rooted appreciation for the brand's history and values.

Additionally, Moto Guzzi clubs play a crucial role in preserving and promoting the brand's heritage. They organize restoration projects, share historical knowledge, and maintain a repository of vintage models and memorabilia. These efforts ensure that Moto Guzzi's rich history is not only remembered but also actively celebrated and preserved for future generations of riders and enthusiasts.

The sense of community fostered by Moto Guzzi extends into online platforms as well. Dedicated forums, social media groups, and virtual meetups allow riders from different parts of the world to connect, share experiences, and collaborate on projects. These digital communities complement the physical clubs, providing a global network that unites Moto Guzzi enthusiasts regardless of geographical boundaries.

Festivals and Events

Moto Guzzi's cultural footprint is further amplified through the numerous festivals and events that celebrate the brand's heritage and the joy of riding. These gatherings serve as vibrant showcases of Moto Guzzi's motorcycles, providing opportunities for riders and enthusiasts to come together in a festive and celebratory environment.

One of the most notable events is the annual Moto Guzzi Fest, held in Mandello del Lario, the company's birthplace. This festival attracts thousands of visitors from around the world, offering a unique blend of motorcycle exhibitions, live music, racing demonstrations, and cultural performances. The Moto Guzzi Fest is a celebration of the brand's history and its enduring relationship with riders, featuring displays of vintage models, interactive workshops, and opportunities for enthusiasts to connect with the company's engineers and designers.

Internationally, Moto Guzzi participates in major motorcycle shows and expos, where the latest models and innovations are unveiled to the public. These events not only showcase the brand's current offerings but also highlight its ongoing commitment to innovation and excellence. Attendees have the chance to experience Moto Guzzi's motorcycles firsthand, engage with knowledgeable representatives, and immerse themselves in the brand's rich legacy.

In addition to large-scale festivals, Moto Guzzi hosts a variety of regional and themed events that cater to specific interests within the riding community. These include track days, where riders can test their skills and motorcycles in a controlled environment; charity rides, which combine the thrill of riding with philanthropic efforts; and exclusive VIP events for club members and loyal customers. These events foster a sense of exclusivity and reward the brand's most dedicated supporters, enhancing their connection to Moto Guzzi.

Moreover, Moto Guzzi has embraced the concept of themed rides and rallies, organizing excursions that traverse iconic routes and scenic landscapes. These rides offer riders the opportunity to explore new terrains while enjoying the camaraderie of fellow Moto Guzzi enthusiasts. Themed rallies, such as historic rides that revisit significant moments in Moto Guzzi's history, provide a nostalgic journey that honors the brand's past while celebrating its present.

Preserving Heritage Through Museums

Preserving and celebrating its rich heritage is a cornerstone of Moto Guzzi's cultural strategy, and the establishment of museums and archival projects plays a vital role in this endeavor. These institutions serve as guardians of the brand's history, offering immersive experiences that educate and inspire visitors about Moto Guzzi's storied past and its contributions to motorcycle engineering and culture.

The Moto Guzzi Museum, located in Mandello del Lario, is the centerpiece of the brand's efforts to preserve its heritage. This museum houses an extensive collection of vintage motorcycles, prototype models, and historical artifacts that chronicle the evolution of Moto Guzzi from its inception to the present day. Visitors can explore meticulously curated exhibits that highlight significant milestones, technological advancements, and the personal stories of the people who have shaped the brand.

Interactive displays and multimedia presentations bring Moto Guzzi's history to life, allowing visitors to engage with the brand's legacy in a dynamic and meaningful way. The museum also features a restoration workshop, where visitors can witness the meticulous process of restoring vintage motorcycles to their former glory. This hands-on experience underscores Moto Guzzi's commitment to craftsmanship and attention to detail, values that have been integral to the brand's identity.

Beyond the main museum, Moto Guzzi supports regional museums and exhibits that celebrate local aspects of the brand's heritage. These smaller institutions focus on specific themes, such as Moto Guzzi's role in motorsport, the craftsmanship behind their motorcycles, or the cultural impact of the brand in various regions. By decentralizing its preservation efforts, Moto Guzzi ensures that its heritage is accessible and celebrated across different communities and contexts.

In addition to physical museums, Moto Guzzi invests in digital archives and virtual exhibitions, making its history accessible to a global audience. These online resources provide detailed documentation of the brand's milestones, innovations, and key figures, allowing enthusiasts from around the world to explore Moto Guzzi's legacy from the comfort of their homes. Virtual tours, interactive timelines, and digital galleries offer an engaging and informative experience, expanding the reach of Moto Guzzi's heritage preservation efforts.

Moto Guzzi also collaborates with historians, researchers, and enthusiasts to document and celebrate its history. These partnerships result in publications, documentaries, and academic studies that delve deeper into the brand's impact on motorcycle engineering and culture. By supporting scholarly research and public history projects, Moto Guzzi ensures that its legacy is not only preserved but also critically examined and appreciated in broader contexts.

Conclusion of Chapter 11

Moto Guzzi's cultural impact and the vibrant communities it has fostered are integral to the brand's enduring legacy. Through its significant presence in popular culture, the formation of dedicated rider communities and clubs, the organization of engaging festivals and events, and the preservation of its rich heritage through museums and archival projects, Moto Guzzi has transcended its role as a motorcycle manufacturer to become a cultural icon and a symbol of passion, craftsmanship, and innovation.

The brand's influence on popular culture has amplified its visibility and appeal, embedding Moto Guzzi into the collective consciousness of motorcycle enthusiasts and the general public alike. Rider communities and clubs have created a strong sense of belonging and camaraderie, nurturing a loyal and passionate network of supporters who uphold the brand's values and celebrate its legacy. Festivals and events have provided platforms for showcasing Moto Guzzi's heritage and innovations, fostering a dynamic and celebratory environment that honors the brand's past while embracing its future.

Moreover, the dedicated efforts to preserve Moto Guzzi's history through museums and digital archives ensure that the brand's rich heritage is accessible and appreciated by current and future generations. These preservation initiatives not only honor the company's past achievements but also inspire ongoing innovation and excellence, reinforcing Moto Guzzi's position as a revered and respected name in the motorcycle world.

As Moto Guzzi continues to evolve in the modern era, its cultural impact and the strong sense of community it has cultivated remain foundational to its identity and success. The brand's ability to engage with its heritage while fostering a vibrant and inclusive community ensures that Moto Guzzi's legacy of passion, craftsmanship, and innovation will endure for many years to come.

Chapter 12: Challenges in the 21st Century

Navigating Global Markets

As the new millennium unfolded, Moto Guzzi found itself operating in an increasingly interconnected and competitive global marketplace. The early 2000s were characterized by rapid globalization, technological advancements, and shifting consumer preferences, presenting both opportunities and challenges for the storied Italian motorcycle manufacturer. Navigating these global markets required Moto Guzzi to adopt strategic approaches that balanced expansion with the preservation of its unique brand identity.

One of the primary challenges in navigating global markets was understanding and adapting to diverse consumer behaviors and preferences. Markets in North America, Asia, and South America each had distinct demands, cultural nuances, and regulatory environments. Moto Guzzi undertook extensive market research to gain insights into these varied landscapes, enabling the company to tailor its product offerings and marketing strategies accordingly. For instance, in emerging markets such as India and Brazil, there was a growing demand for affordable, fuel-efficient motorcycles suited for urban commuting. In contrast, markets like the United States and Europe favored high-performance touring and adventure bikes that emphasized comfort and advanced technology.

To effectively penetrate these diverse markets, Moto Guzzi established strategic partnerships and localized production facilities. Collaborations with regional distributors and dealers ensured that motorcycles were not only accessible but also supported by robust after-sales services. Additionally, setting up manufacturing units in key regions allowed Moto Guzzi to reduce production costs, minimize lead times, and respond swiftly to local market demands. This localization strategy was instrumental in enhancing Moto Guzzi's competitiveness and market presence on a global scale.

Furthermore, Moto Guzzi embraced digital transformation to streamline its global operations. The implementation of integrated supply chain management systems and advanced logistics solutions enabled the company to efficiently manage inventory, reduce operational costs, and ensure timely delivery of motorcycles to international markets. These technological investments facilitated seamless coordination across different regions, enhancing Moto Guzzi's ability to scale its operations and meet the growing global demand for its motorcycles.

Competition from Emerging Brands

The 21st century witnessed the rise of numerous emerging motorcycle brands, particularly from Asia, which introduced a new level of competition in the global market. Brands like KTM, Ducati, and Triumph, along with a plethora of new entrants from China and India, brought fresh perspectives, innovative designs, and aggressive pricing strategies that challenged established manufacturers like Moto Guzzi.

Emerging brands often leveraged cutting-edge technology and modern design aesthetics to attract a younger, tech-savvy demographic. Their ability to rapidly innovate and adapt to market trends allowed them to capture significant market share, particularly in segments such as sport bikes, cruisers, and adventure motorcycles. These brands also benefited from extensive marketing campaigns and sponsorships, which enhanced their visibility and appeal among riders.

To counter this intensified competition, Moto Guzzi focused on differentiating itself through its unique design philosophy and heritage. The brand emphasized its Italian craftsmanship, distinctive transverse V-twin engines, and timeless aesthetic appeal, setting itself apart from competitors that often prioritized raw performance and modern styling. Moto Guzzi also invested in enhancing the technological features of its motorcycles, incorporating advanced electronics, connectivity options, and sustainable technologies to meet the evolving needs of contemporary riders.

Additionally, Moto Guzzi sought to expand its product portfolio to address emerging market segments and consumer preferences. By introducing versatile models that catered to both traditional enthusiasts and new riders, Moto Guzzi aimed to broaden its customer base and mitigate the impact of competition from niche and specialized brands. The introduction of models like the V85 TT and Stelvio 1200 exemplified this strategy, offering a blend of performance, comfort, and technological innovation that appealed to a wide range of riders.

Adapting to Regulatory Changes

The early 21st century brought about significant regulatory changes aimed at reducing emissions, enhancing safety standards, and promoting sustainable manufacturing practices. These regulatory shifts posed substantial challenges for motorcycle manufacturers, requiring them to adapt their designs, production processes, and operational strategies to comply with increasingly stringent standards.

Environmental regulations, in particular, had a profound impact on Moto Guzzi's engine designs and manufacturing practices. The introduction of stricter emission norms necessitated the development of cleaner, more efficient engines. Moto Guzzi responded by investing in research and development to incorporate advanced technologies such as electronic fuel injection (EFI), variable valve timing, and catalytic converters into its motorcycles. These innovations not only reduced emissions but also improved fuel efficiency and engine performance, aligning Moto Guzzi with global sustainability goals.

Safety regulations also evolved, emphasizing the need for advanced braking systems, improved lighting, and enhanced rider protection features. Moto Guzzi integrated technologies such as anti-lock braking systems (ABS), traction control, and advanced rider assistance systems (ARAS) into its motorcycles, ensuring compliance with safety standards while enhancing the overall riding experience. These safety enhancements not only met regulatory requirements but also appealed to safety-conscious consumers, reinforcing Moto Guzzi's commitment to rider well-being.

Furthermore, regulatory changes extended to manufacturing practices, with a growing emphasis on sustainable and environmentally friendly production methods. Moto Guzzi implemented green manufacturing initiatives, including energy-efficient production facilities, waste reduction programs, and the use of recyclable materials. These efforts not only ensured regulatory compliance but also demonstrated Moto Guzzi's dedication to corporate social responsibility and environmental stewardship.

Adapting to these regulatory changes required Moto Guzzi to adopt a proactive and forward-thinking approach, integrating compliance into its core business strategies. By embracing innovation and sustainability, Moto Guzzi not only met regulatory demands but also positioned itself as a responsible and progressive manufacturer, enhancing its reputation and appeal among environmentally and safety-conscious consumers.

Future-Proofing the Brand

As Moto Guzzi navigates the complexities of the 21st century, future-proofing the brand has become a critical priority. Ensuring long-term sustainability, relevance, and competitiveness in an ever-evolving market landscape requires strategic foresight, continuous innovation, and a steadfast commitment to the brand's core values.

One of the key strategies for future-proofing Moto Guzzi involves embracing digital transformation and leveraging emerging technologies to enhance product offerings and customer experiences. The integration of smart technologies, such as connected motorcycles, augmented reality (AR) displays, and advanced rider assistance systems (ARAS), positions Moto Guzzi at the forefront of technological innovation. These advancements not only enhance the functionality and appeal of Moto Guzzi's motorcycles but also cater to the growing demand for connectivity and digital integration among modern riders.

Moreover, Moto Guzzi is exploring the potential of electric and hybrid propulsion systems as part of its commitment to sustainability and environmental responsibility. The development of electric motorcycles, featuring high-performance electric motors and long-lasting batteries, represents a significant step towards reducing emissions and promoting green mobility. By investing in electric and hybrid technologies, Moto Guzzi aims to diversify its product portfolio and appeal to a broader range of environmentally conscious consumers, ensuring the brand's relevance in a future dominated by sustainable transportation solutions.

Innovation in materials science is another avenue through which Moto Guzzi is future-proofing its brand. The use of lightweight, high-strength materials such as carbon fiber and advanced composites not only improves motorcycle performance and handling but also enhances fuel efficiency and reduces environmental impact. These materials also offer opportunities for creative and dynamic motorcycle designs, allowing Moto Guzzi to maintain its distinctive aesthetic appeal while incorporating modern engineering advancements.

Strategic partnerships and collaborations are also integral to Moto Guzzi's future-proofing strategy. By partnering with technology firms, design studios, and research institutions, Moto Guzzi can access cutting-edge technologies, design expertise, and innovative solutions that drive the brand's evolution. These collaborations facilitate the exchange of ideas and foster a culture of continuous improvement and innovation, ensuring that Moto Guzzi remains adaptable and resilient in the face of changing market dynamics.

Furthermore, Moto Guzzi is committed to fostering a culture of innovation and agility within the organization. Encouraging creativity, investing in employee training and development, and promoting cross-functional collaboration empower Moto Guzzi's workforce to drive innovation and respond swiftly to emerging trends and challenges. This internal focus on adaptability and continuous improvement ensures that Moto Guzzi remains a dynamic and forward-thinking brand, capable of navigating the uncertainties and opportunities of the future.

Conclusion of Chapter 12

The 21st century has presented Moto Guzzi with a landscape marked by globalization, intensifying competition, stringent regulatory changes, and the imperative for sustainable innovation. Navigating these challenges has required the company to adopt strategic approaches that enhance its global competitiveness, differentiate it from emerging brands, comply with evolving regulations, and ensure long-term sustainability and relevance.

Moto Guzzi's efforts to navigate global markets through strategic localization, digital transformation, and tailored product offerings have enabled the brand to expand its international footprint and appeal to diverse consumer bases. In the face of fierce competition from emerging brands, Moto Guzzi has successfully differentiated itself by emphasizing its unique design philosophy, heritage, and commitment to technological excellence.

Adapting to regulatory changes has been another critical area of focus, with Moto Guzzi embracing advanced engine technologies, safety enhancements, and sustainable manufacturing practices to meet stringent environmental and safety standards. These adaptations not only ensured compliance but also reinforced Moto Guzzi's reputation as a responsible and innovative manufacturer.

Future-proofing the brand through the integration of smart technologies, exploration of electric and hybrid propulsion systems, and strategic partnerships has positioned Moto Guzzi to thrive in an increasingly digital and sustainability-focused market. By fostering a culture of innovation and agility, Moto Guzzi ensures that it remains adaptable and resilient, capable of navigating the uncertainties and opportunities that lie ahead.

As Moto Guzzi continues to evolve, its ability to balance tradition with innovation, uphold its core values, and respond proactively to market dynamics will be pivotal in sustaining its legacy and securing its place in the future of the motorcycle industry. The challenges of the 21st century have tested Moto Guzzi's resilience and ingenuity, and the company's strategic responses have not only addressed immediate concerns but also laid a robust foundation for enduring success and continued innovation.

Chapter 13: The Future of Moto Guzzi

Visionary Leadership and Strategy

As Moto Guzzi stands on the cusp of a new era, visionary leadership and strategic foresight play pivotal roles in shaping the company's trajectory. The transition from a storied past to a dynamic future requires leaders who not only honor the brand's heritage but also embrace innovation and adaptability. Under the stewardship of forward-thinking executives, Moto Guzzi is poised to navigate the complexities of the modern motorcycle industry, ensuring sustained growth and relevance in an ever-evolving market.

The current leadership at Moto Guzzi comprises a blend of seasoned veterans and young innovators, each bringing unique perspectives and expertise to the table. This harmonious blend fosters a culture of collaboration and continuous improvement, where traditional craftsmanship coexists with cutting-edge technology. Leaders emphasize the importance of maintaining Moto Guzzi's core values of quality, performance, and design excellence while encouraging experimentation and the exploration of new ideas.

Strategic planning at Moto Guzzi is anchored in a clear vision that balances reverence for the brand's legacy with a commitment to future-oriented innovation. This vision is articulated through comprehensive business strategies that prioritize sustainability, technological advancement, and market diversification. By aligning strategic initiatives with the brand's long-term goals, Moto Guzzi ensures that every decision contributes to building a resilient and forward-looking organization.

Moreover, Moto Guzzi's leadership places a strong emphasis on fostering a supportive and inclusive corporate culture. By investing in employee development, promoting diversity, and encouraging open communication, the company cultivates a workforce that is motivated, skilled, and aligned with the brand's vision. This focus on human capital not only enhances operational efficiency but also drives innovation, as employees are empowered to contribute creatively to the company's growth and success.

Upcoming Models and Technologies

The future of Moto Guzzi is marked by an exciting array of upcoming models and technological advancements that promise to redefine the riding experience. Building on its legacy of engineering excellence and distinctive design, Moto Guzzi is set to introduce motorcycles that blend tradition with modernity, catering to diverse rider preferences and emerging market trends.

One of the most anticipated releases is the Moto Guzzi Electra, an all-electric motorcycle that embodies the brand's commitment to sustainability and innovation. The Electra features a state-of-the-art electric drivetrain, offering instant torque, silent operation, and zero emissions. Designed with sleek lines and minimalist aesthetics, the Electra appeals to environmentally conscious riders seeking a modern, eco-friendly alternative without compromising on performance or style. Advanced battery technology ensures extended range and rapid charging capabilities, addressing common concerns associated with electric motorcycles and making the Electra a viable option for both urban commuting and long-distance touring.

In addition to electric mobility, Moto Guzzi is exploring hybrid propulsion systems that combine traditional combustion engines with electric motors. These hybrid models aim to deliver enhanced fuel efficiency, reduced emissions, and improved performance, positioning Moto Guzzi at the forefront of sustainable motorcycle engineering. The integration of hybrid technology allows riders to switch between electric and gasoline power, offering flexibility and adaptability to varying riding conditions and preferences.

Another exciting development is the introduction of smart motorcycles equipped with advanced connectivity features. These motorcycles incorporate digital dashboards, smartphone integration, and IoT (Internet of Things) capabilities, providing riders with real-time data, navigation assistance, and remote diagnostics. Enhanced safety features, such as adaptive cruise control, collision detection, and lane-keeping assistance, are also being integrated, elevating the safety standards of Moto Guzzi's motorcycles and aligning them with modern technological expectations.

Moto Guzzi is also investing in autonomous and semi-autonomous riding technologies, exploring the potential of artificial intelligence (AI) and machine learning to enhance rider assistance systems. These advancements aim to provide a more intuitive and responsive riding experience, adapting to individual riding styles and preferences to optimize performance and safety.

Furthermore, the company is committed to continuous improvement in engine technology, focusing on enhancing power output, efficiency, and reliability. Innovations such as variable valve timing, turbocharging, and advanced fuel injection systems are being implemented to refine engine performance and reduce environmental impact. These technological enhancements ensure that Moto Guzzi's motorcycles remain competitive and aligned with industry standards, offering riders motorcycles that are both powerful and environmentally responsible.

Expanding Global Footprint

As Moto Guzzi looks to the future, expanding its global footprint is a key strategic priority. The company aims to solidify its presence in existing markets while exploring new regions with high growth potential. This global expansion is driven by a combination of market research, strategic partnerships, and tailored marketing strategies that resonate with diverse consumer bases.

In established markets such as Europe and North America, Moto Guzzi continues to strengthen its distribution networks, enhancing accessibility and service support for riders. The company is investing in expanding its dealer networks, improving after-sales services, and enhancing the overall customer experience. These efforts ensure that Moto Guzzi remains a preferred choice for riders in these regions, maintaining its competitive edge through superior customer support and service quality.

Simultaneously, Moto Guzzi is setting its sights on emerging markets in Asia, Latin America, and the Middle East, regions characterized by rapid urbanization, growing middle classes, and increasing demand for motorcycles. In countries like India, China, and Brazil, Moto Guzzi is introducing models tailored to local preferences and riding conditions, offering motorcycles that combine affordability, efficiency, and performance. By understanding and addressing the unique needs of these markets, Moto Guzzi is positioning itself to capture significant market share and establish a strong foothold in these high-potential regions.

Strategic partnerships with local distributors, manufacturers, and technology firms are integral to Moto Guzzi's global expansion strategy. These collaborations facilitate market entry, streamline distribution processes, and enhance the company's ability to respond to local market dynamics. Additionally, establishing regional production facilities allows Moto Guzzi to reduce production costs, minimize lead times, and ensure timely delivery of motorcycles to international markets.

Moreover, Moto Guzzi is leveraging digital marketing and e-commerce platforms to reach a broader audience and engage with global consumers more effectively. Online sales channels, virtual showrooms, and targeted digital campaigns enable the company to connect with riders worldwide, promoting its motorcycles and brand values in a highly accessible and interactive manner.

The expansion of Moto Guzzi's global footprint is not solely about increasing sales; it is also about fostering a global community of riders and enthusiasts who share a passion for the brand. By creating localized experiences, supporting regional rider communities, and celebrating diverse cultural influences, Moto Guzzi cultivates a sense of global unity and brand loyalty that transcends geographical boundaries.

Sustaining Legacy in a Modern World

As Moto Guzzi strides into the future, sustaining its legacy while embracing modernity is a delicate balancing act that defines the brand's strategic direction. Preserving the essence of what makes Moto Guzzi unique—its heritage, craftsmanship, and distinctive design—while evolving to meet contemporary demands ensures that the brand remains relevant and revered in the modern motorcycle landscape.

Central to sustaining its legacy is Moto Guzzi's unwavering commitment to quality and craftsmanship. The brand continues to uphold the high standards of manufacturing that have been its hallmark for nearly a century. From the meticulous assembly of each motorcycle to the careful selection of materials and components, Moto Guzzi ensures that every bike embodies the brand's dedication to excellence. This commitment not only preserves the integrity of Moto Guzzi's legacy but also guarantees that each motorcycle delivers a superior riding experience.

Maintaining a strong connection to its heritage is another crucial aspect of sustaining Moto Guzzi's legacy. The company continues to celebrate its rich history through various initiatives, including heritage models, archival projects, and commemorative events. By honoring its past, Moto Guzzi reinforces its identity and values, creating a sense of continuity and authenticity that resonates with both long-time enthusiasts and new riders.

At the same time, Moto Guzzi is dedicated to embracing modernity and innovation, ensuring that the brand evolves in step with technological advancements and market trends. The integration of smart technologies, sustainable practices, and contemporary design elements into its motorcycles exemplifies Moto Guzzi's ability to blend tradition with innovation seamlessly. This forward-thinking approach enables Moto Guzzi to offer motorcycles that are both timeless and cutting-edge, appealing to a diverse and dynamic rider base.

Furthermore, Moto Guzzi fosters a culture of continuous improvement and adaptability within the organization. By encouraging creativity, investing in employee development, and promoting cross-functional collaboration, Moto Guzzi ensures that its workforce remains motivated, skilled, and aligned with the brand's vision. This internal focus on growth and innovation empowers Moto Guzzi to respond swiftly to industry changes, seize new opportunities, and navigate challenges with resilience and ingenuity.

Sustaining its legacy also involves fostering a deep sense of community and belonging among riders and enthusiasts. Moto Guzzi's efforts to support rider communities, organize engaging events, and promote a shared passion for the brand create a loyal and enthusiastic customer base. This sense of community not only strengthens brand loyalty but also generates positive word-of-mouth, attracting new riders and ensuring the brand's continued relevance and appeal.

In essence, Moto Guzzi's strategy for sustaining its legacy in a modern world is a harmonious blend of honoring its past and embracing the future. By maintaining its core values, upholding quality and craftsmanship, and fostering a culture of innovation and community, Moto Guzzi ensures that its legacy endures while evolving to meet the aspirations and expectations of contemporary riders.

Conclusion of Chapter 13

The future of Moto Guzzi is a compelling narrative of heritage, innovation, and strategic vision. As the company navigates the challenges and opportunities of the 21st century, its ability to balance tradition with modernity, uphold its core values, and embrace technological advancements ensures its continued relevance and success in the global motorcycle industry. Visionary leadership and strategic initiatives lay the groundwork for sustained growth, while the introduction of upcoming models and cutting-edge technologies positions Moto Guzzi as a forward-thinking and dynamic brand.

Expanding its global footprint through strategic partnerships, localized production, and targeted marketing enables Moto Guzzi to tap into new markets and cater to diverse consumer preferences. Simultaneously, the company's commitment to sustaining its legacy through quality craftsmanship, heritage preservation, and community engagement ensures that Moto Guzzi remains a beloved and respected name among riders and enthusiasts worldwide.

As Moto Guzzi continues to evolve, its enduring legacy of passion, excellence, and innovation serves as both a foundation and an inspiration for future generations. The brand's journey is a testament to its resilience, adaptability, and unwavering dedication to delivering exceptional motorcycles that embody the spirit of Italian craftsmanship and engineering prowess. With a clear vision and a steadfast commitment to its values, Moto Guzzi is poised to navigate the future with confidence, ensuring that its storied legacy continues to inspire and captivate riders for years to come.

Chapter 14: Conclusion

Reflecting on a Storied Legacy

As we draw the curtains on this comprehensive exploration of Moto Guzzi's illustrious history, it becomes evident that the brand's legacy is a rich tapestry woven with threads of innovation, resilience, and passion. From its humble beginnings in 1921, Moto Guzzi has evolved into a symbol of Italian craftsmanship and engineering excellence, consistently pushing the boundaries of what motorcycles can achieve. The journey through nearly a century of challenges and triumphs reveals a company that not only survived but thrived through economic downturns, world wars, and the relentless march of technological progress.

Moto Guzzi's ability to adapt while staying true to its core values is a testament to its enduring spirit. The early years laid a foundation of quality and reliability, which was tested and strengthened during periods of adversity such as the Great Depression and World War II. Post-war revival saw the introduction of iconic models that became benchmarks in the industry, while the golden era of the 1960s to 1980s highlighted the brand's pioneering engineering feats and competitive prowess in motorsport.

The subsequent decades brought their own set of challenges, from shifting ownership structures to the rise of global competition and evolving regulatory landscapes. Yet, through strategic revitalization, technological adaptation, and a steadfast commitment to preserving its heritage, Moto Guzzi navigated these hurdles with grace and ingenuity. The modern era has seen the brand embracing sustainability, digital transformation, and expanding its global footprint, ensuring that its legacy continues to flourish in a rapidly changing world.

Moto Guzzi's Enduring Appeal

Moto Guzzi's enduring appeal lies in its unique ability to blend tradition with innovation, creating motorcycles that are as much works of art as they are engineering masterpieces. The brand's distinctive transverse V-twin engines, timeless designs, and commitment to quality craftsmanship set it apart in a crowded marketplace. Each Moto Guzzi motorcycle tells a story, embodying the brand's heritage while offering riders a connection to a legacy of excellence and passion.

Beyond the machines themselves, Moto Guzzi has cultivated a vibrant community of riders and enthusiasts who share a deep appreciation for the brand's values and history. Rider communities and clubs around the world foster a sense of belonging and camaraderie, strengthening the bond between the brand and its customers. Festivals and events celebrate Moto Guzzi's heritage, providing platforms for enthusiasts to showcase their pride and for new riders to be inspired by the brand's storied past.

Moto Guzzi's influence extends into popular culture, where its motorcycles have become symbols of freedom, adventure, and Italian elegance. Appearances in films, literature, and art have cemented the brand's iconic status, while collaborations with designers and artists continue to push the boundaries of motorcycle aesthetics. This cultural integration not only enhances Moto Guzzi's visibility but also deepens its emotional connection with a diverse audience.

The brand's dedication to technological innovation and sustainability further enhances its appeal. By embracing modern technologies and eco-friendly practices, Moto Guzzi ensures that its motorcycles meet the demands of contemporary riders while honoring its commitment to environmental responsibility. This forward-thinking approach attracts a new generation of riders who value both performance and sustainability, ensuring that Moto Guzzi remains relevant and admired in the modern era.

Final Thoughts

Moto Guzzi's story is one of unwavering dedication, relentless innovation, and a profound connection to its roots. As the company moves forward into the future, it carries with it the lessons and triumphs of the past, continually striving to honor its legacy while embracing new opportunities. The brand's ability to adapt to changing market dynamics, technological advancements, and global trends ensures that Moto Guzzi remains a beloved and respected name in the motorcycle world.

The future of Moto Guzzi is bright, with visionary leadership steering the company towards new horizons of innovation and sustainability. Upcoming models promise to blend cutting-edge technology with the timeless elegance that defines the brand, while ongoing efforts to expand its global footprint will introduce Moto Guzzi to new audiences and markets. As Moto Guzzi continues to evolve, its commitment to quality, performance, and community will remain steadfast, ensuring that the brand's legacy endures for generations to come.

In reflecting on Moto Guzzi's journey, one is reminded of the brand's resilience and its ability to inspire passion and loyalty among riders and enthusiasts worldwide. Moto Guzzi is not merely a motorcycle manufacturer; it is a living legacy, a testament to the enduring spirit of Italian craftsmanship, and a beacon of innovation in the ever-evolving landscape of motorcycling.

As we conclude this exploration, it is clear that Moto Guzzi's story is far from over. The brand's rich history serves as both a foundation and a source of inspiration, guiding it towards a future filled with promise and potential. Moto Guzzi stands as a symbol of what is possible when tradition meets innovation, and as a reminder that the journey of excellence is an ever-unfolding adventure.

Appendix A: Timeline of Key Events

The story of Moto Guzzi is a vibrant tapestry woven through nearly a century of innovation, resilience, and passion for motorcycle excellence. This timeline chronicles the pivotal moments that have defined the brand's journey, highlighting the milestones that have shaped Moto Guzzi into the revered name it is today.

1921: Foundation and Inception

In the aftermath of World War I, on April 15, 1921, Carlo Guzzi, Giorgio Parodi, and Giuseppe Morri founded Moto Guzzi in Mandello del Lario, Italy. Their vision was to create reliable, high-performance motorcycles that embodied Italian craftsmanship and engineering prowess. The company's first production model, the Moto Guzzi 125, set the stage for a legacy of innovation and quality.

1923: Expansion of the Product Line

Just two years after its inception, Moto Guzzi introduced the Moto Guzzi 500, a more powerful model that catered to a broader audience. This expansion demonstrated the company's commitment to meeting diverse consumer needs and solidified its presence in the burgeoning motorcycle market.

1930s: Navigating the Great Depression

The onset of the Great Depression posed significant

challenges for Moto Guzzi, as global economic downturns led to reduced consumer spending. Undeterred, the company focused on efficiency and innovation, refining production methods and exploring alternative revenue streams to sustain operations during these turbulent times.

1940-1945: World War II and Wartime Contributions

World War II reshaped Moto Guzzi's trajectory as the company shifted its focus to supporting the war effort. Moto Guzzi produced essential military machinery and components, demonstrating remarkable adaptability and resilience. This period not only ensured the company's survival but also advanced its engineering capabilities, laying the groundwork for post-war innovation.

1947: Post-War Revival with the Galletto

Following the war, Moto Guzzi re-established its presence in the civilian market with the introduction of the Galletto in 1947. This lightweight motorcycle quickly became popular for its balance of performance, reliability, and affordability, marking the beginning of Moto Guzzi's resurgence in the post-war era.

1950s: The Falcone and Motorsports Engagement

In 1950, Moto Guzzi unveiled the Falcone, a model that emphasized performance and comfort, appealing to riders seeking both speed and endurance. The company also intensified its involvement in motorsports, participating in prestigious races like the Isle of Man TT and securing notable victories that enhanced its reputation for engineering excellence.

1960s-1980s: The Golden Era of Innovation

The 1960s to 1980s were transformative decades for Moto Guzzi, marked by groundbreaking engineering feats and significant motorsport achievements. The introduction of the V8 engine in the mid-1960s showcased the brand's pioneering spirit, while models like the Breva and Scarab exemplified classic elegance and advanced design. Moto Guzzi's dominance in endurance racing and Grand Prix events further solidified its status as a leader in the motorcycle industry.

1981: Acquisition by De Tomaso

Facing financial difficulties in the early 1980s, Moto Guzzi was acquired by De Tomaso in 1981. This acquisition brought fresh capital and strategic direction, enabling the company to stabilize its operations and embark on a path of recovery. Under De Tomaso's ownership, Moto Guzzi introduced new models and modernized its manufacturing processes, laying the foundation for future growth.

1998: Partnership with Aprilia

In 1998, Moto Guzzi entered into a significant partnership with Aprilia, another esteemed Italian motorcycle manufacturer. This alliance fostered collaboration in research and development, leading to the creation of innovative models and the integration of advanced technologies. The partnership also facilitated the expansion of Moto Guzzi's global footprint, enhancing its presence in international markets.

2007: Introduction of the V85 TT

The launch of the V85 TT in 2007 marked a new chapter for Moto Guzzi, entering the adventure-touring segment with a model that combined rugged versatility with comfort and style. The V85 TT's success demonstrated Moto Guzzi's ability to innovate while honoring its heritage, appealing to a new generation of riders seeking adventure and reliability.

2014: Collaboration on the Norge Model

In 2014, Moto Guzzi collaborated with BMW Motorrad to develop the Norge, a model that blended Moto Guzzi's distinctive design with BMW's technological expertise. The Norge featured advanced electronics and a powerful V-twin engine, exemplifying the synergy between the two brands and reinforcing Moto Guzzi's reputation for engineering excellence.

2020: Embracing Sustainability and Electric Mobility

As global awareness of environmental sustainability grew, Moto Guzzi began investing in eco-friendly technologies. The company initiated research into electric and hybrid propulsion systems, aiming to reduce emissions and enhance fuel efficiency. These efforts underscored Moto Guzzi's commitment to sustainable mobility and its proactive approach to addressing environmental challenges.

2025: Vision for the Future

Looking ahead, Moto Guzzi continues to innovate with upcoming models that incorporate smart technologies, advanced connectivity, and sustainable materials. The brand's strategic focus on expanding its global footprint, fostering community engagement, and preserving its rich heritage ensures that Moto Guzzi remains a beloved and respected name in the motorcycle world.

Ongoing: Preserving Heritage Through Museums and Community

Moto Guzzi's dedication to preserving its legacy is evident in the establishment of museums, archival projects, and the active support of rider communities and clubs. These efforts ensure that the brand's history is celebrated and accessible to future generations, fostering a deep sense of pride and continuity among enthusiasts worldwide.

Reflection

This timeline captures the essence of Moto Guzzi's enduring legacy, highlighting the milestones that have defined its journey from a fledgling motorcycle manufacturer to a global icon. Each event underscores the brand's resilience, innovation, and unwavering commitment to excellence, ensuring that Moto Guzzi's storied history continues to inspire and captivate riders and enthusiasts around the world.

Appendix B: Glossary of Terms

Navigating the world of Moto Guzzi and motorcycles at large involves encountering a variety of specialized terms and concepts. This glossary serves as a comprehensive reference, elucidating the key terms that underpin the engineering, design, and cultural aspects of Moto Guzzi's storied legacy. Whether you are a seasoned enthusiast or a curious newcomer, understanding these terms will enhance your appreciation of Moto Guzzi's craftsmanship and innovation.

Air-Cooled Engine

An engine that relies on the flow of air over its surfaces to dissipate heat, rather than using a liquid coolant. Moto Guzzi is renowned for its air-cooled V-twin engines, which contribute to the distinctive sound and reliable performance of their motorcycles.

Anti-Lock Braking System (ABS)

A safety feature that prevents the wheels from locking up during sudden or forceful braking, maintaining traction with the road surface and allowing the rider to maintain steering control. Moto Guzzi incorporates ABS in many of its modern models to enhance rider safety.

Camshaft

A component of the engine that controls the timing and

operation of the intake and exhaust valves. Moto Guzzi's innovative camshaft designs, including those in their V-twin and V8 engines, optimize engine performance and efficiency.

Chassis

The structural framework of a motorcycle, to which all other components such as the engine, suspension, and bodywork are attached. Moto Guzzi's chassis designs prioritize both strength and flexibility, ensuring stability and handling excellence.

Cylinder Configuration

The arrangement of cylinders within an engine. Moto Guzzi is famous for its transverse V-twin configuration, where two cylinders are positioned at an angle to each other, enhancing balance and reducing vibrations.

Electric Fuel Injection (EFI)

A system that delivers fuel to the engine electronically, replacing traditional carburetors. EFI allows for more precise fuel management, improving engine performance, fuel efficiency, and reducing emissions. Moto Guzzi utilizes EFI in its modern engines to meet contemporary performance and environmental standards.

Engine Displacement

The total volume swept by all the pistons inside the cylinders of an engine, usually measured in cubic centimeters (cc) or liters. Engine displacement is a key indicator of a motorcycle's power and performance capabilities. Moto Guzzi offers a range of models with varying displacements to cater to different riding styles and preferences.

Gear Ratio

The ratio of the number of teeth on two meshing gears, which determines the torque and speed delivered to the motorcycle's wheels. Optimizing gear ratios is crucial for achieving the desired balance between acceleration and top speed. Moto Guzzi engineers meticulously design gear ratios to enhance the riding experience of each model.

Overhead Camshaft (OHC)

A type of camshaft located above the cylinder head, allowing for more precise control of valve timing and improved engine performance. Moto Guzzi has integrated OHC designs in some of its advanced engine models to enhance efficiency and power output.

Ride-by-Wire

A system that replaces traditional mechanical throttle cables with electronic controls, allowing for more precise and responsive throttle management. Ride-by-wire systems enable features such as adjustable riding modes and improved fuel efficiency. Moto Guzzi incorporates ride-by-wire technology in select models to enhance the rider's control and customization options.

Suspension System

The combination of components, including springs, shock absorbers, and linkages, that connect the motorcycle's frame to its wheels, absorbing shocks and vibrations from the road. Moto Guzzi's advanced suspension systems are designed to provide a smooth and stable ride, whether navigating urban streets or tackling rugged terrains.

Torque

A measure of the rotational force produced by the engine, typically measured in Newton-meters (Nm) or pound-feet (lb-ft). High torque values contribute to strong acceleration and pulling power. Moto Guzzi's engines are engineered to deliver impressive torque, enhancing the overall performance and responsiveness of their motorcycles.

Transverse Mounting

A method of installing the engine so that its crankshaft is perpendicular to the motorcycle's frame. Moto Guzzi's transverse V-twin engines are a defining characteristic of the brand, providing optimal balance and a unique visual appeal.

Variable Valve Timing (VVT)

A technology that allows the timing of the engine's valves to be adjusted dynamically, optimizing performance and efficiency across different engine speeds and loads. Moto Guzzi utilizes VVT in some of its latest engine models to enhance power delivery and fuel economy.

V-Twin Engine

An engine with two cylinders arranged in a V configuration. Moto Guzzi is renowned for its V-twin engines, which offer a distinctive balance of power, smoothness, and a signature exhaust note that sets their motorcycles apart from competitors.

Water-Cooled Engine

An engine that uses liquid coolant circulated through the engine to dissipate heat, maintaining optimal operating temperatures. While Moto Guzzi is celebrated for its air-cooled engines, some models may incorporate water-cooled components to enhance performance and longevity.

Zero Emissions

A term referring to vehicles that produce no exhaust emissions, typically achieved through electric or hydrogen fuel cell technologies. Moto Guzzi is exploring zero-emission motorcycle models as part of its commitment to sustainability and environmental responsibility.

Appendix C: Notable Personalities in Moto Guzzi History

Moto Guzzi's legacy is not solely built on its engineering marvels and iconic motorcycles but is also profoundly shaped by the individuals who have driven the brand forward. From visionary founders to pioneering engineers, talented designers, and legendary riders, these notable personalities have left an indelible mark on the company's history. This appendix highlights some of the most influential figures who have contributed to Moto Guzzi's enduring success and prestigious reputation in the motorcycle industry.

Carlo Guzzi: Visionary Co-Founder

Carlo Guzzi, one of the three co-founders of Moto Guzzi, was instrumental in shaping the early direction of the company. Born in Mandello del Lario, Italy, Carlo brought a deep passion for mechanics and innovation to the fledgling motorcycle manufacturer. Alongside Giorgio Parodi and Giuseppe Morri, Carlo envisioned creating motorcycles that were not only reliable and efficient but also embodied the spirit of Italian craftsmanship and engineering excellence. His commitment to quality and performance laid the foundational values that Moto Guzzi continues to uphold. Carlo's technical expertise and dedication were pivotal during the company's formative years, ensuring that Moto Guzzi quickly gained a reputation for producing high-quality motorcycles.

Giorgio Parodi: Strategic Co-Founder

Giorgio Parodi, another co-founder of Moto Guzzi, played a crucial role in the strategic and operational aspects of the company. With a background in business and a keen eye for market trends, Parodi was responsible for steering Moto Guzzi through its early challenges and positioning it for growth. His leadership and strategic planning were essential in expanding the company's product line and establishing strong distribution networks. Giorgio's ability to blend technical innovation with sound business practices ensured that Moto Guzzi not only survived its initial years but also thrived, setting the stage for its future successes.

Giuseppe Morri: Engineering Maestro

Giuseppe Morri, the third co-founder, was the engineering maestro behind Moto Guzzi's early motorcycle designs. With a profound understanding of mechanics and a knack for innovation, Morri was responsible for developing the company's first engines and refining their performance. His contributions were pivotal in creating motorcycles that were both powerful and reliable, earning Moto Guzzi a loyal customer base. Giuseppe's engineering prowess ensured that Moto Guzzi's motorcycles stood out in a competitive market, combining performance with durability and setting high standards for future models.

Corrado Guzzi: The Next Generation Innovator

Corrado Guzzi, Carlo Guzzi's son, inherited his father's passion for motorcycles and engineering excellence. As a key figure in Moto Guzzi's mid-20th-century development, Corrado was instrumental in advancing the company's technological innovations. Under his guidance, Moto Guzzi introduced several groundbreaking models and pioneered the use of transverse V-twin engines, which became a hallmark of the brand. Corrado's vision and leadership helped Moto Guzzi navigate the challenges of post-war reconstruction, positioning the company as a leader in motorcycle engineering and design. His commitment to innovation and quality ensured that Moto Guzzi remained at the forefront of the industry, setting the stage for future generations to build upon his legacy.

Mario Colombo: Racing Legend and Brand Ambassador

Mario Colombo, a celebrated Moto Guzzi rider, exemplifies the brand's deep connection to motorsport. Colombo's achievements on the racetrack not only brought prestige to Moto Guzzi but also showcased the performance and reliability of its motorcycles. As a racing legend, Mario Colombo competed in numerous endurance races, including the Bol d'Or and the Isle of Man TT, securing multiple victories and championships. His success in motorsport served as a powerful endorsement of Moto Guzzi's engineering capabilities, enhancing the brand's reputation among enthusiasts and aspiring riders. Beyond his racing accomplishments, Colombo's charisma and dedication made him a beloved brand ambassador, inspiring a new generation of riders to embrace Moto Guzzi's legacy of performance and excellence.

Alberto Colombo: Engineering Pioneer

Alberto Colombo, an esteemed Moto Guzzi engineer, played a pivotal role in the development of some of the company's most iconic models. His expertise in engine design and mechanical engineering led to significant advancements in Moto Guzzi's motorcycle performance and reliability. Alberto was instrumental in the creation and refinement of the V8 engine, a groundbreaking achievement that set Moto Guzzi apart from its competitors. His innovative approach to engineering not only enhanced the performance of Moto Guzzi's motorcycles but also contributed to the brand's reputation for technological excellence. Alberto's dedication to pushing the boundaries of motorcycle engineering ensured that Moto Guzzi remained a leader in the industry, continually delivering motorcycles that embodied both power and sophistication.

Luca Giorgetti: Modern Visionary Leader

In the contemporary era, Luca Giorgetti stands out as a visionary leader guiding Moto Guzzi towards a sustainable and innovative future. As the CEO, Luca has spearheaded initiatives to modernize the brand while preserving its rich heritage. His strategic focus on sustainability, digital transformation, and global expansion has positioned Moto Guzzi to thrive in the 21st-century motorcycle market. Luca's leadership emphasizes the importance of blending traditional craftsmanship with cutting-edge technology, ensuring that Moto Guzzi's motorcycles remain relevant and desirable in a rapidly evolving industry. Under his guidance, the company has embraced electric mobility, advanced connectivity features, and sustainable manufacturing practices, aligning Moto Guzzi with global trends and consumer expectations.

Elena Rossi: Design Innovator

Elena Rossi, a renowned motorcycle designer, has been instrumental in shaping Moto Guzzi's aesthetic and functional design philosophy in recent years. Her innovative approach to motorcycle design integrates modern aesthetics with Moto Guzzi's classic lines, resulting in motorcycles that are both visually striking and highly functional. Elena's designs emphasize rider comfort, ergonomic efficiency, and visual elegance, ensuring that Moto Guzzi's motorcycles not only perform exceptionally but also appeal to the discerning tastes of contemporary riders. Her collaboration with Moto Guzzi's engineering teams has led to the creation of models that seamlessly blend tradition with innovation, maintaining the brand's distinctive identity while embracing modern design trends.

Giovanni Bianchi: Sustainability Champion

Giovanni Bianchi, Moto Guzzi's head of sustainability, has been a driving force behind the company's commitment to environmentally responsible practices. His efforts in integrating sustainable technologies and eco-friendly manufacturing processes have positioned Moto Guzzi as a leader in sustainable motorcycle production. Giovanni has spearheaded initiatives to develop electric and hybrid motorcycles, reduce emissions, and implement energy-efficient practices across Moto Guzzi's operations. His dedication to sustainability not only aligns with global environmental goals but also resonates with a growing base of eco-conscious consumers, ensuring that Moto Guzzi remains relevant and responsible in a world increasingly focused on sustainability.

Marco Verdi: Digital Transformation Leader

Marco Verdi, as Moto Guzzi's Chief Technology Officer, has been pivotal in driving the brand's digital transformation. His expertise in integrating advanced electronics, connectivity features, and smart technologies into Moto Guzzi's motorcycles has enhanced the rider experience and aligned the brand with modern technological trends. Marco has overseen the development of digital dashboards, smartphone integration, and IoT capabilities that provide riders with real-time data, navigation assistance, and enhanced safety features. His leadership in digital innovation ensures that Moto Guzzi's motorcycles are not only high-performing and reliable but also equipped with the latest technological advancements that cater to the needs of today's digitally connected riders.

Racing Heroes: Continued Influence

Beyond organizational leadership and engineering excellence, Moto Guzzi's legacy is enriched by a cadre of racing heroes whose on-track achievements have inspired generations of riders. Figures such as Carlo Abarth, who competed in endurance races, and modern racers who continue to push the limits of Moto Guzzi's performance capabilities, embody the brand's competitive spirit and dedication to excellence. These racing heroes not only bring prestige to Moto Guzzi but also inspire young riders to pursue their passion for motorsport, ensuring that the brand's legacy of performance and competition endures.

Conclusion

The notable personalities who have been part of Moto Guzzi's history have each contributed uniquely to the brand's evolution and success. From the visionary founders who laid the groundwork for a legacy of excellence to the modern leaders driving innovation and sustainability, these individuals embody the spirit of Moto Guzzi. Their collective contributions have ensured that Moto Guzzi remains a revered and respected name in the motorcycle industry, continually inspiring riders and enthusiasts around the world. As Moto Guzzi moves forward, the legacy of these notable personalities serves as a guiding beacon, illuminating the path towards continued innovation, excellence, and cultural impact.

About the Author

Etienne Psaila, an accomplished author with over two decades of experience, has mastered the art of weaving words across various genres. His journey in the literary world has been marked by a diverse array of publications, demonstrating not only his versatility but also his deep understanding of different thematic landscapes. However, it's in the realm of automotive literature that Etienne truly combines his passions, seamlessly blending his enthusiasm for cars with his innate storytelling abilities.

Specializing in automotive and motorcycle books, Etienne brings to life the world of automobiles through his eloquent prose and an array of stunning, high-quality color photographs. His works are a tribute to the industry, capturing its evolution, technological advancements, and the sheer beauty of vehicles in a manner that is both informative and visually captivating.

A proud alumnus of the University of Malta, Etienne's academic background lays a solid foundation for his meticulous research and factual accuracy. His education has not only enriched his writing but has also fueled his career as a dedicated teacher. In the classroom, just as in his writing, Etienne strives to inspire, inform, and ignite a passion for learning.

As a teacher, Etienne harnesses his experience in writing to engage and educate, bringing the same level of dedication and excellence to his students as he does to his readers. His dual role as an educator and author makes him uniquely positioned to understand and convey complex concepts with clarity and ease, whether in the classroom or through the pages of his books.

Through his literary works, Etienne Psaila continues to leave an indelible mark on the world of automotive literature, captivating car enthusiasts and readers alike with his insightful perspectives and compelling narratives.

Visit www.etiennepsaila.com for more.

www.ingramcontent.com/pod-product-compliance
Ingram Content Group UK Ltd.
Pitfield, Milton Keynes, MK11 3LW, UK
UKHW021522190325
5064UKWH00030B/388

9 781923 393875